Dictionary of Oil and Gas Production

Clifford Jones

Published by
Whittles Publishing,
Dunbeath,
Caithness KW6 6EG,
Scotland, UK

www.whittlespublishing.com

ISBN 978-184995-047-3

Distributed in the US
by CRC Press

Cover photo: Øyvind Hagen/Statoil

Printed and bound in England

Contents

Dedicated to:

Dr. Hyun-key Kim Hogarth
A woman of erudition and humanity

Preface

This is the third in my series of dictionaries, following those on Energy and Fuels (co-authored by Nigel Russell) and Fire Protection Engineering. It is a satisfying experience for the compiler of a specialist dictionary to organise the information and to link the entries to provide a balanced coverage of value to a wide range of readers.

Oil and natural gas are in historical terms quite recent: the inevitable benchmark date is the drilling of the Drake Well in Titusville PA in 1859. That was just a few years before King George V, grandfather of the present Queen, was born, so although the beginning of the oil industry is not within living memory it is within the range of oral tradition. Most of us know something about our families a couple of generations ago and the circumstances of their lives[1]. Two to three generations – a short enough interval in the sweep of history – have seen oil and gas develop from a new industry to one upon which every living soul is dependent and which has a very strong effect indeed on politics and world finance. Over thirty-five years ago I heard the journalist John Pilger say on the BBC: 'A man can live without oil: he cannot live without food'. If for 'A man' we substitute 'The world's population' (probably not a misrepresentation of what Pilger was saying) that statement might need reconsidering today.

Who can deny that oil and natural gas have brought incalculable benefits to the world? In the pre-industrial era the same would have been said of agriculture and, as with any asset, abuse entails loss. The term 'superannuation' is now just about synonymous with 'post-retirement income', but the concept of superannuation goes deeper than that. In times long before the discovery of oil and gas when people depended on 'the land' for their livelihood a field was 'superannuated' when it was in need of lying fallow for one or more seasons to allow nutriment replenishment and prevent erosion. The dust bowls of the 1930s are a sad reminder of the effects of neglecting to apply 'superannuation' in that sense.

I have gone into this detail of difficulties in the agricultural milieu because, over months of close study and evaluation of oil production in the 20th and early 21st Centuries, I have developed a deep consciousness that social responsibility *has* prevailed. I believe that any informed individual making a similar study could not but reach the same conclusion. This impression, encouraging though it is, is not in any way intended to be the thrust of this volume. It is simply that it has arisen in my own mind so often over the research and writing periods that I want to include it in this preface.

Notwithstanding its being a 'dictionary', this volume does contain much close engineering reasoning and indeed some related calculations. I hope that students, professionals in the oil and gas industry and policymakers will benefit from it. I shall be delighted to receive feedback from readers. As always, my thanks go to Whittles Publishing.

J.C. Jones, Aberdeen

Foreword

It is a pleasure to be able to write a foreword for another of Clifford Jones' dictionaries, though the name dictionary does not fully describe the little gems of information and explanation which accompany each listing. Each profession has its own combination of jargon and 'things everybody knows' which can be very hard for the newcomer to pick up. This mini-encyclopaedia will, we are sure, be a great help.

However, the oil and gas industry is particularly diverse, ranging from drilling to end use of fuels, so that even people in one part of the business will not be aware of some items in other areas. This book will also be of great help to experienced people wishing to broaden their knowledge or anticipating a move within the industry – an increasingly likely event in today's flexible labour market and global pattern of industry.

It is truly impressive how much information has been condensed into these pages: geological terms, information about oil and gas fields, equipment and even in many cases methods of calculation explained with an example. We may both claim some years of professional experience in different areas of the oil and gas industry, and were therefore pleased to find many things we did know, but slightly surprised at the amount we did not. We shall certainly make use of it in the future.

Prof Vida N. Sharifi, CEng FInstE
Dr Martin J. Pitt, CEng FIChemE
Chemical & Biological Engineering, University of Sheffield, UK

Abqaiq, crude oil stabilisation at

This location in Saudi Arabia is the scene of the world's largest plant for crude oil stabilisation, with a capacity of 7 million barrels a day. Off-gas from the stabilisation process is refined for use and ethane is totally separated from it. There is also hydrogen sulphide removal from the oil which, in the stabilised and sweetened state in which it emerges from the process, is taken to one of three refineries. Two of these are in Saudi Arabia and the other, a **Bapco** facility, is in Bahrain.

Aceh

In North Sumatra, Indonesia. There is abundant natural gas in the region but activity was, in the early 2000s, strongly affected by war there. It was also affected by the tsunamis in 2004. Gas from one block offshore Aceh supplies part of the requirement of the **Arun LNG plant**. The block is operated by Mobil and has a **pay** of 335 m. There is a condensate refinery in Aceh, although this takes only a fraction of the total condensate, most of which is sold at a spot price. Across the industry, spot price sale of condensate has become prevalent only relatively recently and regional, e.g. Texan, rather than national or international spot prices apply. The author has no details of the basis of spot prices for condensate to hand, but one would expect intuititively that the density of the condensate either as produced or after stabilisation would be the determining factor, more probably the latter. (*See also* **Gajah Baru Field; Sulawesi.**)

Acergy Condor

Pipe laying vessel, built in 1982. Maintenance over the years has included SPS Overlay treatment in November 2006. In September 2010 it commenced a contract for pipe laying at water depths up to 2000 m offshore Brazil. Pipe laying vessels of the Acergy fleet have used both **S-Lay** and **J-Lay** but neither will apply to the current activity of Acergy Condor as it is flexible pipe that is being laid. Such pipe obviously owes its flexibility to its structure, which comprises a non-metallic outer layer ('sheath') with spirally wound metal spring wire along its inside surface. About 75% of such pipe internationally is made by Technip. It can be recovered and re-used, making its use attractive in decommissioning, as has indeed been the case in the North Sea. It has been used at the **Etame Field**, which has a life expectancy of only a few years, and so its retrieval will be an option. Its flexibility means that a pipeline on the seabed and a riser to the production facility can be linked without a joint. (*See also* **Casablanca Field**; **Conkouati**; **Lula Field**.)

Acid fracturing pumper

Offshore acid stimulation of a well is done from a well intervention vessel. Onshore it is done from a pump mounted on a chassis and cab, and an example of such a device is the NOV Rolligon Acid Fracturing Pump unit. Its specifications are that its pumping power is 1350 b.h.p. to 1500 b.h.p. for acid treatment, and it can provide part of the power required for hydraulic fracture. Here the semantics issue apropos of 'acid' and 'fracture' explained in an endnote in this volume arises. The unit is known as an acid fracturing unit even when (as will be so for most of the time) it is only admitting acid for chemical treatment of the formation. One can understand why the unit under discussion cannot on its own perform a hydraulic fracture by comparing it with one expressly built for hydraulic fracture, with pumping fluid containing **proppant**. One such, manufactured by CAT GmBH in Germany, has a pump capable of 2250 b.h.p., much more powerful than that of the NOV Rolligon. Even so, the CAT unit can operate below full power for simple acid treatment of onshore wells. Two points can be made from this comparison. First, that flexibility is needed in the design of such a unit so that it is not restricted to one sort of operation. Secondly, such flexibility is a factor in the blurring of terminology. Is the driver of the CAT unit going to give it a different name depending on whether the destination is a hydraulic fracture job or an acid stimulation job?! (*See* **Bluell formation ND**.)

Acoustic excitation

(*See* **Sonication, well stimulation by**)

ADGAS Plant

In Abu Dhabi, now having three LNG trains with a combined production capacity of 7.6 million tonnes per annum. This averages to 633 000 tonnes per month, although there have been times when the monthly production has been greater than 750 000 tonnes, significantly exceeding nameplate capacity, which arguably needed revision after the condenser on one of the refrigeration units, with propane as refrigerant, was modified in the direction of enhanced productivity. (*See also* **Bahia de Bizkaia Regasification Plant**; **Bukhoosh Field**; **Formates** (endnote); **Umm Shaif Field**.)

ADIP® process

Developed by Shell, a means of removing hydrogen sulphide from natural gas. Where high selectivity for hydrogen sulphide is required the absorbing agent is **MDEA** whereas **DIPA** is used if there is also to be removal of carbon dioxide. The process is carried out in a packed adsorption column. Temperatures are up to 60 °C and pressures up to 150 bar. Having regard to the fact that the basis of removal is absorption not chemical reaction, the absorber is readily regenerated after exit and can be readmitted. Production of elemental sulphur by the Claus processs is an option at this stage. There are about 500 ADIP® process units in service worldwide at the present time.

Agbami Field

In the Niger Delta, producing oil since 2004. The water depth is a little over 1450 m and the operator is Chevron. During appraisal drilling was to 4780 m and a total **pay**, across several zones, of 163 m was identified. Drilling of the production wells was by **Deepwater Discovery**. The oil is of density 45 degrees API (800 kg m^{-3}), making it 'light', and is also low in sulphur, making it 'sweet'. The benchmark price which applies to oil from offshore Nigeria is of course the OPEC basket. Production is by an **FPSO** also called Agbami which can store over 2 million barrels and is expected to remain in service at the field for 20 years. Associated gas is reinjected. In Q3 2009 the field was producing 0.25 million barrels per day. (*See also* **Akepo Field**; **Glomar Explorer**.)

Aging oil pipelines, use of drag reducing agent at

When a pipeline is becoming elderly this can be factored into risk analysis and de-rating imposed accordingly. For example, if the walls of a gas pipeline are showing evidence of corrosion the original design stress can be reduced for calculation purposes. The equivalent for passage of *liquid hydrocarbons* is that the closer to being streamline the flow is the smaller are the stresses on the internal pipe wall. A **drag reducing agent** can be used to achieve such an effect, enabling an elderly pipeline to remain in service as long as provision is made for such flow control. **Turboflo™** has been so used. (*See* **Marmul to Nimr pipeline**.)

AHD

Along hole departure, a term used in well drilling. It can be understood by reference to the diagram below which represents a well and shows the true vertical depth (**TVD**) and its relationship to the AHD. Clearly, the AHD is given by:

$$AHD = TVD \tan\Theta$$

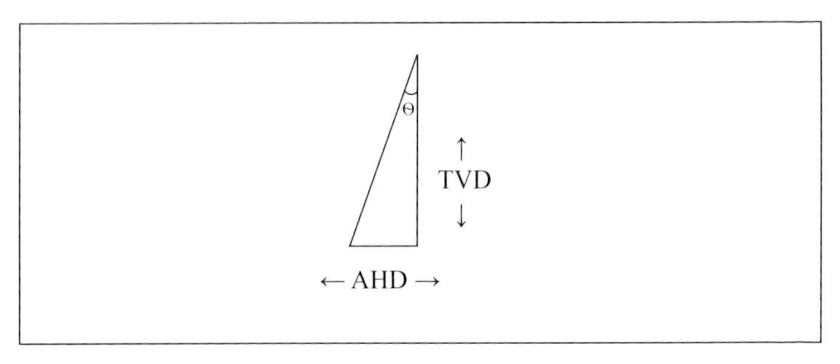

When there is no **directional drilling** Θ = 0, tanΘ = 0 and the AHD is zero. In **extended reach drilling** the AHD-to-TVD ratio is an important number called the ERD ratio. At the BD-04A well in the **Al Shaheen Field** offshore Qatar a record ERD ratio of 10.485 was set, giving Θ the value 85 degrees. A value so close to that for a right angle is only realisable for long drilling distances. At BD-04A the AHD was 11.6 km and the **TVD** 1.1 km giving:

$$\Theta = \tan^{-1}(11.6/1.1) = 85°$$

consistent with the statement a few lines above. The AHD at BD-04A is also said to be a record and provides a basis for comparison with the **Chayvo Field**. The company carrying out the drilling at the BD-04A well was Schlumberger. The ERD figure for the well can be compared, for example, with those for wells at the West Sak Oil field on the North Slope of Alaska where ERD values of 1.37 (Θ = 54 degrees) and of 2.04 (Θ = 64 degrees) have been reported. (See also **Armada**; **Boulton Field**; **DDI**; **Tortuosity, in well drilling**; **Sapele well**; **TVD**.)

Airth

In central Scotland, site for **CBM** production, which began as a pilot scheme in 2004. There has been limited production, a peak of 42 000 cubic metres per day having been reported in 2009[2]. At the time of writing production is suspended and awaiting further development. The developer is Composite Energy, who have drilled four horizontal wells at Airth. It is believed that about 13 cubic metres of CBM per tonne of coal in situ can be realised. (*See* **Wales, exploration for CBM in**.)

Aitingen

Oil field in Germany operated by Wintershall in association with ExxonMobil and Production Deutschland GmbH. There is gas and condensate additional to the oil. Well depths are greater than 1200 m and daily oil production is about 600 barrels per day, very moderate without making the field non-viable. It has been producing at a rate never exceeding 800 barrels per day for over 30 years: the fact that the oil is light – 860 kg m^{-3} – is a plus. The attitude which seems to prevail in Germany is that it is precisely *because* production from its own oil fields falls well short of the national demand for oil that they must be husbanded with all possible care and diligence. (*See* **Emlichheim**.)

Akepo Field

Oil field in the Niger Delta, water depth 8 m. Exploratory drilling in 1993 revealed a **pay** of 32.5 m of oil and subsequent sidetracks revealed 45 m of oil pay and 33.5 m of gas pay. The **Lloyd Noble** was deployed in drilling an appraisal well. It is expected that oil from the field will be taken to an existing **FSO** by agreement with its owner.

Al Ghallan

Jack-up drilling rig built in 1976 in Scotland and for a very long time in service in the Middle East with the National Drilling Company (NDC), Abu Dhabi. Its original specifications were such that it could drill in up to 41 m of water to well depths of greater than 6000 m. Upgrading took place in 2008, being carried out under the direction of NDC by Lamprell. The modernised Al Ghallan has a new cantilever-and-jack structure. The latter increases the sea depth that can be accessed by 14 m and the former increases the range of horizontal positioning of the drill by 3.7 m, so the drilling envelope has been significantly enlarged by the modification. (*See* **Offshore Freedom**.)

Al Jalila Field

Newly discovered (2010) oil field offshore the UAE, commencement of production expected before the end of 2011. Little information has been released at the time of writing, beyond that it lies east of the **Rashid Field**.

Al Jurf Field

Field offshore Libya[3], water depth about 90 m. There are ten production wells which are connected to an **FPSO** having a storage capacity of 0.9 million barrels. The operator is Total. (*See* **Farwah**.)

Al Shaheen Field

Oil field with associated gas offshore Qatar, discovered in 1992. Initial assessment was that the field was marginal, and developer Maersk drew on experience of such fields in the North Sea. Production required horizontal drilling, and by 1996 there was a **jack-up** production platform in 62.5 m of water supported by **spud cans**. Hydrocarbon from this was taken to a loading buoy with SPM. Expansion followed, and over the period 1998 to 2001 40 new production wells were drilled and 20 water injection wells. Further wells followed, one of which, as recorded in the entry for **AHD**, set a record for extended reach drilling. There are now about 30 platforms at the field. Two 'sister' double hulled **FSOs** – FSO Africa and FSO Asia – entered service there in early 2010 and in July the same year a cumulative production of one billion barrels was attained.

Albury

Small onshore gas field in the UK, operated by Star Energy who are a subsidiary of Petronas. It was discovered in 1987 and production began in 1994. The gas is not transported by pipeline, but burnt close to the scene of production to generate electricity, which goes to the national grid. Also operated by Star Energy are a number of other small fields in the UK, including Humbly Grove in Hampshire. Discovered in 1980, the field has been producing since 1985; its production rate has over that period been of the order of 1000 barrels per day. The oil is on the sour side (0.35% sulphur). It is distributed under an agreement by Esso. (*See* **Welton**.)

All Seal

Lost circulation material manufactured by the Texas based Alliance Drilling Fluids. It consists of finely ground cellulose which will swell on entry into a 'spider web formation'. Similar products from the same company use cottonseed husks or cedar-wood fibres. Another Alliance Drilling Fluids product is **My-Lo-Jel**.

Alliance pipeline

This conveys gas from British Columbia and Alberta to the Chicago area, and was constructed more recently than **GTN** which also carries gas from Canada to the US. GTN 'travels' south east and Alliance south west. Alliance carries over 30 million cubic metres of gas per day. There are no receiving facilities on the Canadian side of the pipeline: all the gas goes to Chicago apart from small amounts offloaded in Iowa and in North Dakota for use in making methanol. Other such diversions on the US side of the pipeline are 'on the drawing board'. Alliance has seven **compressor stations** with a combined power of 218 400 b.h.p. (*See* **Vector pipeline**.)

Altimira LNG terminal

In Mexico, receiving its first shipment in 2006. The companies involved are Shell, Total and Mitsui. The first shipment referred to was from Nigeria, and an agreement has been entered into whereby LNG from Australia's North West Shelf will be taken by tanker to Altimira. The facility has storage tanks and **ORVs** and will supply LNG for uses including electricity generation. (*See* **Costa Azul LNG terminal**.)

Altmark Field

In Germany, before the recent sharp decline the second largest onshore gas field (next to Groningen) in western Europe. Occupying an area of 1000 square kilometres, it has been producing since 1969 from well depths of about 3500 m. There were once 450 active wells at Altmark, but now only 150, and since 1985, the year of peak production, the annual yield of gas has declined from 12 billion cubic metres to less than one billion cubic metres. 'Enhanced gas recovery' is being discussed but is a long way from implementation.

AMA®–324

Biocide from Kemira, applications of which include **SRB** and **APB** control in oil reservoirs and pipelines. The active ingredient is tetrahydro-3,5dimethyl-2H-1,3,5-thiadiazine-2-thione.

Amelia platform

(*See* **Guendalina Field**.)

Amine process

A means of removing hydrogen sulphide from natural gas. It is analogous to **glycol dehydration** in that the H_2S does not chemically react with the amine but is absorbed by it. The most widely used amines for the purpose are **MEA** and **DEA** and each also absorbs carbon dioxide. Efflux gas from an amine process can be further treated by the Claus process, which partially oxidises the H_2S to elemental sulphur, a saleable product. At the Kharg Island Gas Gathering and Natural Gas Liquids Recovery Project, 1200 tonnes per day of sulphur are so produced for sale. (*See also* **ADIP® process**; **Kharg Island, Iran**.)

Amirante

Semi-submersible owned and operated by Transocean and currently in use at the **Appaloosa Field**. Built in Finland and having undergone two major modifications, it can function in water depths up to 1067 m and can drill to depths of up to 7620 m. Its **moonpool** is 19.7 feet by 19.5 feet and its **drawworks** are rated at 3000 h.p. (*See* **Longhorn Field**.)

Ampa Field

Offshore Brunei, like the **Champion Field** containing many (more than 400) reservoirs some of which contain oil and some non-associated gas[4].

The latter predominates (about 60%) in the products from the field. The gas, from 56 gas wells within the complex of reservoirs, goes to **Lumut** for conversion to LNG. There are 164 oil production wells.

AMPS/AAm copolymer

Copolymer of 2-acrylamido-2-methylpropyl sulphonic acid and acrylic acid having found application in water based drilling muds. Baker Hughes manufacture such a product having the trade name Pyro-Trol® and this has been used both in seawater and in freshwater muds. Its presence influences how effectively the mud lubricates the drill bit (its 'lubricity') as well as determining the viscosity. Its continued functioning can be ensured by inclusion of an agent such as **SSMA**.

Andrew Field

Relatively small field in the UK sector of the North Sea, water depth 115 m. There is a single production platform, and production since commencement in 1996 has been up to 80 000 barrels per day. The **pay** of oil is 58 m and that of gas 60 m. (*See* **Easywell**.)

Angel Field

Gas and condensate offshore north-west Australia. The water depth is 80 m and there are three production wells. There is a **fixed platform** from which the gas is transferred through pipes to a pipeline structure from a larger platform ('North Rankin'). The Angel Field produces roughly 20 million cubic metres per day of gas, which is destined for conversion to liquefied natural gas (LNG) and export to countries including Japan. The daily condensate yield is 50 000 barrels. By a calculation the same as that given in detail for the **Yttergryta Field** the weight ratio of condensate to gas is 0.43. Production at about these levels is expected until 2017. (*See* **Condensate stripping**.)

Angot Field

Small oil field in Afghanistan, discovered in 1959. Between then and 2006 fifteen production wells came into operation each producing 50 barrels per day, that is, 750 barrels per day in all. The formation has porosities in the range 17.5% to 28% and permeabilities in the range 48 to 800 millidarcy. The present owner is Ghazanfar Neft Gas which operates petrol stations around the country (supplied, of course, with petrol from foreign oil). In order for the hoped-for development at Angot to take place many more wells will need to be drilled and pipelines laid, and this will not be possible without foreign investment.

Annular blowout preventer

This consists of an elastomer device, colloquially termed a 'rubber doughnut'. In normal well operation the hole at the centre of it forms part of the well bore, but when an annular **BOP** is activated the elastomer is compressed so that it spreads in the radial direction and closes off the well bore. The shape and configuration of the elastomer structure are such as to maximise spread when compression occurs. The annular blowout preventer is simpler to construct and to install than the **ram-type blowout preventer** and has the advantage that it can be used across a range of well bore diameters. **Nitrile rubber** is often used as the elastomer.

Anticor

Emulsion breaker manufactured in Malaysia by Scobi, who have several patented chemicals and blend them to make an emulsion breaker suitable for a particular application. Like other manufacturers of such products, Scobi send technicians to 'the field' to develop and test a specially made-up emulsion breaker. Anticor so adjusted to requirement has been used in Nigeria as well as at more than one European oil field.

APB

Acid producing bacteria, bacteria other than and additional to **SRB**, that cause corrosion in pipelines. In general their role in corrosion is smaller than that of SRB. APB include bacteria of the genus *Thiobacillus*. APB and SRB often act synergistically. Sulphate ions at the stage where they are reduced by SRB to sulphide (S^{2-}) are in salt solution at or close to neutral pH. Certain APB can oxidise the sulphide back to sulphate with accompanying hydrogen ion formation:

$$S^{2-} + 4H_2O \rightarrow 8H^+ + SO_4^{2-} + 8e$$

and clearly the products will form a sulphuric acid solution. pH values of three if not lower can arise in a pipeline in this way. The term MIC – microbially induced corrosion – means corrosion due to APB and SRB individually or together.

Appaloosa Field

In the Gulf of Mexico, water depth 853 m. It is currently undergoing development and the operator is Eni, who discovered oil at the field in 2008. Drilling at that stage was by the semi-submersible **GSF Celtic Sea**. At the present time **Amirante** is being deployed there in further drilling. Production wells and **water injection** wells at Appaloosa will be tied back to the Corral (formerly Crystal) platform. (See **Longhorn Field**.)

Apsara Field

In the Gulf of Thailand in Cambodian waters. If the aspiration to have produced oil there by the end of 2012 is fulfilled it will be the first 'Cambodian oil'. The developer is Chevron and there were 'shows' of oil and gas at seven of the nine exploration wells drilled. Proposals are for a production platform of a capacity of *up to* 25 000 barrels per day, oil from which will go to a permanently moored **FSO**. It is estimated that the existing well structure to which the platform is connected will 'reach' a total of 10 million barrels of crude oil. Cumulative production of this amount will not signify depletion: extension by way of new wells of sidetracking existing ones will follow.

Aqua-Clear®

Well rehabilitation product from Halliburton having the nature of a polymer dispersant. It is also effective in reducing the viscosity of drilling muds. (*See* **BMR™**.)

Aquifer water drive

A.k.a. aquifer sweep or aquifer influx. (*See* **Harding Field**.)

Aral Sea

'Inland sea' within Uzbekistan, salinity of the order of 100 gram per litre and observed to have been rising sharply since the 1960s when it was the centre of a major fishing industry. There is natural gas under the Aral Sea, and in 2010 half a million cubic metres were produced there. It is planned to connect the field with the **Central Asia-China gas pipeline**.

Arcasolve™

Chemical agent for filter cake removal manufactured by Chemusorb in the UK. It is acidic, enabling it to react with inorganics in the filter cake, and also capable of breaking down polymers which might have been present in mud to which the rock formation had been exposed. At the **Ormen Lange** gas field in the North Sea additional wells have recently been drilled, and Arcasolve™ was applied to them in order to repair 'drilling mud damage'. In well **completion** closure of the **FIV** is intended to protect the contents of the gravel pack from contamination but where this has not totally worked Arcasolve™ can be used to clean the gravel. Once, at a well at a field offshore South America, Arcasolve™ used in the well casing was found to be amongst the 'contaminants' of the gravel pack. The simple expedient of allowing time for it to dissolve the other contaminants of the gravel pack was effective and the **PI** of the well was as hoped for. Being acidic, Arcasolve™ has also found application in well stimulation.

Argentina, unconventional gas plays in

Total have involvement with unconventional gas in a number of countries including Argentina. The company is active in tight gas production at Aguada Pichana in the Neuquén Basin where permeabilities are low. In the words of a conference abstract written by one of Total's own officers in Argentina[5], this places Aguada Pichana in the 'tight gas reservoir domain'. The field had previously produced conventionally, and hydraulic fracture was applied to access the tight gas. Twenty-five new wells were drilled and nine of the previously existing ones had workovers. Production at the field so developed is 5 million cubic metres per day. Total distinguish the gas at Aguada Pichana from 'shale gas' also from the Neuquén Basin, including that at the Cerro Partido play. Total, operating with Argentina's own YPF, currently hold six **shale natural gas** licences in that part of Argentina. (*See* **Sulige Field**.)

Armada

Fixed platform in the UK sector of the North Sea receiving from three fields: Fleming, Drake and Hawkins, collectively known as the Armada development. The water depth is about 90 m and **extended reach drilling** was applied in the unitization of the fields. For example, the well from Fleming to the platform was of total length 7286 m, with a **TVD** of 2743 m. This gives a deviation of:

$$\cos^{-1}(2743/7286) = 68°$$

The corresponding calculation for Drake is:

$$\cos^{-1}(2743/5782) = 62°$$

where the same TVD value has been used for each, an assumption justified when a diagrammatic cross-section through the Armada development is examined. Neither well, of course, is a straight line: deviation for the well from Fleming is *up to* 77° and that for the well from Drake is *up to* 63°, a value very close to that obtained above which involves a number of approximations. Peak annual production at Armada is 115 million cubic metres of gas and 0.8 million tonnes of condensate.

Arthit Field

Condensate field in the Gulf of Thailand of which the operator is PTT Exploration and Production, with Chevron and Mitsui also having holdings. The water depth is 80 m. Initial production began in March 2008 and by June of that year production was 10.5 million cubic feet of natural gas per day, and 19 800 barrels of condensate per day. Drilling of the production wells was by **jack-up** rigs including one from the Ensco stable. There is a central processing platform at the field for operations including **condensate stripping**, followed by pipeline transfer of the gas and liquid products.

Arun LNG Plant

In the **Aceh** province of Sumatra, receiving gas from the field of the same name and from other fields. The plant now has six LNG trains, and the LNG is exported to South Korea and (in smaller amounts) to Japan. There are reports that, although its continued operation is not in question, the Arun LNG Plant is in decline at the time of writing (November 2011) and that production for 2011 will be well down on the 2010 figure. This is linked to reduced production in the Arun field itself. (*See* **Bontang LNG plant**.)

Åsgard pipeline

This carries gas from the Åsgard field to the **Kårstø processing plant** near Stavanger, a distance of 707 km. **Dual-diameter pigging** is practised there. (*See* **TTRD**.)

Asphalt, in drilling muds

It is clear from other entries that additives for filtration control can be humic or polymeric in nature. Another substance having found such application is asphalt, composed of high molecular weight (>1000) compounds found in crude oil, called asphaltenes. An example is the Halliburton product AK-70®.

Associated gas, value of

Where gas occurs with crude oil in the same reservoir it might not be economic to collect it, in which case it is simply flared. Some oil-producing countries have been criticised for flaring large amounts of gas which, it is claimed, could have been diverted to fuel use and policy is to collect gas rather than flare it. Nowadays simple flaring takes up carbon credits, and this makes collection of smaller amounts more feasible than formerly. With the **Prirazlomnoye Field** as an example, the value in heat terms of the oil and gas simultaneously produced are calculated in the box below. An interested reader can easily perform the same calculation for other fields.

The field produces 14 000 barrels per day of oil and 10^6 m^3 per day of gas[6].

A barrel of oil on burning releases 6 GJ of heat, therefore

a day's oil from the Prirazlomnoye Field is capable of releasing 8×10^{13} J (80 TJ) of heat.

The calorific value of natural gas is about 37 MJ m^{-3} (the m^3 referred to 1 bar, 288K). A day's gas from the Prirazlomnoye Field can therefore release 4×10^{13} J (40 TJ) of heat, half that released by the oil.

Where the topside of the production facility is not set up for gas handling, gas can be reinjected into the well, as is currently taking place at many sites of hydrocarbon production. It should be remembered that 'joule-to-dollar conversion' is different for oil and for natural gas, so the factor of two in the above example does not apply on a monetary basis. To convert on such a basis one would need to use simultaneous benchmark prices for each, e.g., West Texas Intermediate (oil) and Henry Hub (gas): simultaneous, because the ratio of one to the other varies and is affected by the price of oil. The ratio of oil to gas price per unit energy was unprecedentedly erratic during the Gulf Coast Hurricanes. (*See also* **Bengkulu Basin**, **Troll Field**.)

Astra

Jack-up rig having recently seen service, on behalf of Dragon Oil, offshore Turkmenistan. Built in Japan in 1983, it can operate in water depths up to 36 m and can drill to depths approaching 5000 m. (*See also* **Dzheitune Field**; **Iran Khazar**.)

Atlantic LNG Plant

In Trinidad. The good fortune of Trinidad and Tobago in discovering large amounts of natural gas within its maritime boundaries required development along one or more of three lines: connection to infrastructure in nearby Venezuela for pipeline transfer; conversion to methanol; and conversion to LNG. The second has predominated, with BP on the methane upstream (gas production) side and Methanex producing, via syngas, the methanol for tanker export. There is however interest in the third, LNG production, with British Gas as one participant. An LNG plant at Port Fortin, Trinidad, is on a site which was previously a refinery. Commencing operations in 2000, the facility now has four LNG trains. The LNG is taken to the south-eastern USA for conversion back to gas and onward pipeline distribution. As for the first possibility mentioned – co-operation with Venezuela – two points of interest can be made. One is that in 2003 proposals whereby gas should be sent *to* Trinidad from Venezuela for processing did not come to fulfilment. Secondly, an agreement was recently made between the two nations in relation to the Loran-Manatee gas field, through which the maritime border between Trinidad and Tobago (T&T) and Venezuela passes. Venezuela obtained by far the larger amount. The view has been expressed that T&T had no incentive to develop Loran-Manatee until there was evidence of incipient depletion of its own gas fields. (*See also* **Cove Point MD**; **Grain LNG Terminal**; **Kiskadee Field**; **North Adriatic LNG Terminal**; **St John; New Brunswick**.)

ATOT

Arctic Tandem Offloading Terminal, a means of transferring oil from a production facility in very icy waters. It uses an **OIB** with turret mooring to the facility at which the oil is being produced. Moored 'in tandem' is a **shuttle tanker** built for such conditions, often called an Arctic tanker. The linked OIB–shuttle tanker will then be operated in one of two ways. One is *close tow operation* in which there is no movement of the OIB and the shuttle tanker relative to one another during oil transfer. Clearly this will be the approach taken when the ice is so abundant that there is insufficient sea movement to affect the union of the OIB and the shuttle tanker. Where ice is not as prevalent *distant tow operation* is used and during the offloading the OIB and the shuttle tanker have to be separated by 60–80 m to prevent collision of one with the other. In either case there is net movement of the OIB–shuttle tanker pair, as even highly consolidated ice moves with the wind and with sea currents beneath the ice. Once offloading is completed the OIB can be replenished and the shuttle tanker can, after removal of the mooring lines, take the oil to a terminal. The reader will have appreciated that the primary purpose of an OIB is the same as that of a loading buoy with SPM under conventional conditions. An OIB does however have the means to cut and to grind ice. Note that when ATOT is carried out the production facility might itself be onshore. (*See also* **Icebreaker Sakhalin**; **Molikpaq**.)

Attapulgite

Mineral compound of formula:

$$(Mg,Al)_2Si_4O_{10}(OH)\cdot 4(H_2O)$$

used in drilling muds. It gives a lower **clay yield** than **bentonite**, values around 7 $m^3\ t^{-1}$ being expected. Like bentonite and **sepiolite**, attapulgite can be treated to make it organophilic and in that form it can be used in an oil based mud.

Attic oil

Oil at the top of a reservoir not removed in the course of regular production, therefore remaining when the field starts to deplete. At the Dunlin Field off Shetland in the North Sea there has been major recovery of attic oil by a novel (and award-winning) procedure involving in situ generation of an inert gas from reagents admitted to the parts of the reservoir containing attic oil. (*See* **Lake Maracaibo**.)

Atwood Aurora

Jack-up rig, entering service in 2008. It can drill to subsea depths of 9000 m. Once in situ it is supported by **spud cans** of area 140 m^2. The **derrick** can take a load of about 700 tonnes. Its **BOP** assembly comprises one **annular blowout preventer** and one **ram-type blowout preventer**. Mooring is by four 1½ inch wire ropes attached to anchors. (*See* **Papyrus-1X**.)

Atwood Hunter

Semi-submersible rig having recently been in service in the **Tamar Field**. The rig can operate in water depths up to 1500 m and can drill to depths of 8500 m. Its **derrick** can take a load of 540 tonnes and its **drawworks** are rated at 3000 h.p. Since completion of its appraisal in undertaking well drilling duties at Tamar, it has been taken to waters off west Africa.

Atyrau

In Kazakhstan, on the Caspian coast, terminus of two systems which provide for export of oil: the **Caspian pipeline consortium** from which oil is taken to Russia and the **Kazakhstan-China oil pipeline** from which oil is taken to the border with China. There has been oil refining at Atyrau since 1945. (*See also* **Karachaganak Field**; **Kemerkol project**.)

Audacia

Pipe laying vessel, more recent than **Solitaire** but not as big. Audacia was an adaptation of an existing vessel, not a new build. It has **dynamic positioning** and can lay pipe in the o.d. range 2 inches to 60 inches. Its initial service, in 2007, was in the North Sea.

Auk Field

Exploration and (subsequently) production licences are allocated on the basis of blocks, and what an 'explorer' finds within the site allocated can vary in geological content. A good example is the Auk Field in the North Sea, part of the reservoir of which is sandstone and part dolomite. The permeability of a formation does not of course depend solely on its chemical nature but also on the conditions of structure and deposition. It is stated in another entry in this volume that the sandstone reservoir at Alwyn North has a permeability approaching 1 darcy (1000 md). The sandstone part of the Auk Field has a permeability of 5 md, which is very low. The dolomite part has a permeability of 53 md. The respective **porosity** values are 13% and 19%, both in the range commonly observed and higher than in the **Buchan Field**, also sandstone. It is only the *permeability* of the sandstone in the Auk Field that makes it anomalous, but not so much so as to preclude production. The water depth there is 82 m, and oil production began in 1975. By the time Talisman acquired the field in 2010 it had produced about 200 million barrels of crude oil. (*See also* **Angot Field**; **Burghley Field**; **Constitution/Ticonderoga Field**; **Fulmar Field**; **Tern Field**; **Vesterled Pipeline**.)

Australia, shale natural gas in

Characteristically, favourable conventional gas prices recently have delayed shale gas exploration in Australia (even though there has been **CBM** production in Queensland for some time). A play in the Gippsland region of Victoria (for generations a source of low-rank coal for electricity production, currently to as great an extent as ever) has been identified for future hydraulic fracture to access tight gas.

AVD™

Active vibration damper. The AVD™ contains a magnetorheological fluid, that is, a fluid whose viscosity can be changed and controlled by application of a magnetic field. The viscosity can therefore be changed according to the thrust at the drill bit. Both drill bit life expectancy and **ROP** are improved as a result.

Ayazli Field

Gas field in the Black Sea, offshore Turkey. Production, from two wells, began in 2008. The production platform there has a tripod base. Once this had been installed, the topside, having been made in Sweden, was towed to the Ayazli Field and lifted on to the tripod support structure, following which the two wells were tied back to the platform.

Az Zubair Field

Oil field in Iraq, undergoing redevelopment by Eni (Italy), Occidental (US) KOGAS (S. Korea) and the **Missan Oil Company** (Iraq). The redevelopment began in 2009, at which time production from the field was 0.2 million barrels per day. For every rise of 10% from the 2009 production level $2 per barrel is payable to the group of companies and the goals set are 1 million barrels per day in 2015 and production at that level for a further seven years. Eni have engaged Halliburton for certain tasks, including well acidising.

Azeri-Chirag-Gunashli Field

Oil field in the Caspian Sea, producing over one million barrels per day.

There are several **fixed platforms** each in 120 m of water. They include three **PDQ** platforms three of which use **water injection**. The associated gas is partly collected, partly used on site, and partly reinjected.

AziTrak™

MWD instrument from Baker Hughes. It detects changes in the geological formation and boundaries including those between water and oil. It works on the principle of measurement of the resistivity ('specific resistance': usual units ohm cm) of the formation. It can anticipate the drill bit by about 5 m in detecting an interface of sedimentary rock (a.k.a. shale, particles having become consolidated in mud or clay) and non-sedimentary. (*See* **Sureshot™**.)

Other books by the same author

Dictionary of Fire Protection Engineering

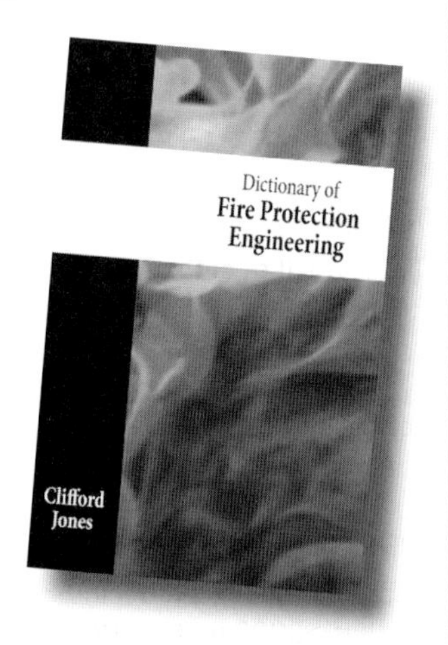

- Up-to-date information on all major aspects of fire protection engineering
- Compiled by a recognized authority on the subject
- Contains a broad yet detailed coverage of the major aspects of fire engineering

ISBN 978-1904445-86-9   129 × 198mm
304pp   softback   £16.99

Dictionary of Energy and Fuels

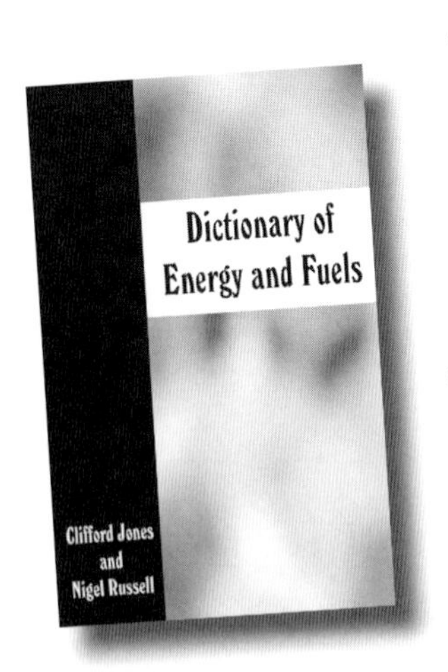

- '...a staggeringly wide range of definitions is on offer ... a useful dictionary in a compact format ... which should satisfy those requiring a one-stop shop of information on a wide range of matters relating to fuel and energy...' ***Fuel***
- '...a compendium of useful information ... a broad spectrum of topics is included in this short but fact-filled volume ... will serve as a very useful reference...' ***Journal of Hazardous Materials***

ISBN 978-1904445-44-9   129 × 198 mm   376pp   softback   £16.99

B

Babbage

Gas field in the southern part of the UK sector of the North Sea having entered production in 2010. Dana Petroleum have a major interest. The water depth is 42 m and there is significant condensate. There was hydraulic fracture during development of the field. There are three production wells and Phase II of the development involves the drilling of two more. Production aimed for ultimately will be 2 million cubic metres per day. (*See* **Dimlington**.)

Bagaja gas condensate field

In Turkmenistan, producing gas and condensate since 2006. The well depth is >3200 metres. Bagaja is said to be the largest 'condensate field' in Turkmenistan. There is an LNG train there and the condensate is taken 37 km by pipeline to the Seydi oil refinery and processed into products for marketing within Turkmenistan. (*See* **Kollsnes**.)

Bahia de Bizkaia Regasification Plant

In Spain, converting LNG carried in tankers to natural gas. Its joint owners include BP. The natural gas produced is diverted to electricity production and the electricity utility itself has a stake in the regasification plant. The plant has facilities for **SCV**. It receives LNG for sources including the **ADGAS Plant** where in fact its very first shipment came from.

Bahrain Field

A.k.a. the Awali Field, onshore field, crude from which goes to a refinery in Bahrain that also receives crude from Saudi Arabia. Its average production is 35 000 barrels per day. The field is undergoing redevelopment in which **Bapco** and Occidental are both involved. This redevelopment involves the drilling of hundreds of new wells, many of them horizontal. The target is a doubling of the field's oil production after 15 years, and a rise to more than 45 000 barrels of oil per day by as soon as 2015. Not only is there associated gas, but in one region of the geologically diverse Awali Field there is a reservoir of non-associated gas. This is in the Khuff formation, which expands into other countries in the region including the UAE, Qatar, Oman and Saudi Arabia. Obviously the non-associated gas from the Awali Field is being factored into the redevelopment programme. There are two **glycol dehydration** units in service to process the non-associated gas. Each has a rating of just under 2 million cubic metres per day. (*See also* **Ampa Field**; **Dukhan Field**; **Umm Shaif Field**; **Yibal Field**.)

Balder

FPU built in Singapore for service in the Norwegian sector of the North Sea on behalf of Esso. It is based at an oil field of the same name (Balder Field) where the water depth is 175 m.

Bangladesh, production of natural gas by Chevron in

Chevron have three natural gas fields in Bangladesh. These are presented in tabular form below.

Name of field	Details
Moulvibazar[7]	Producing since 2005, a few months before the merger of Chevron and Unocal. 2010 production 1.6 million cubic metres per day. Condensate bought by the Bangladesh Petroleum Corporation and refined.
Bibiyana	Discovered in 1998 and producing since 2007. 2010 production was 20 million cubic metres per day. A contract was put in place in 2004 whereby the gas is purchased from Chevron by the state-owned Petrobangla.
Jalalabad	Producing 4.5 million cubic metres of gas per day. Condensate used to make automotive fuel.

The condensate produced at Moulvibazar is taken by road from the field to the refinery. Bibyana, now the largest natural gas field in Bangladesh, was actually discovered by Occidental. The sum of the gas production figures given in the table is:

$$(1.6 + 20 + 4.5) \text{ million cubic metres per day} = 26.1 \text{ million cubic metres per day}$$

and this amounts to a little under half of the natural gas requirement of Bangladesh, with the condensate as a very significant bonus.

Bapco

Bahrain Petroleum Company. (*See also* **Bahrain Field**; **Abqaiq, crude oil stabilisation at**.)

BARADRIL-MAG™

Chemical agent from Halliburton for inclusion in drill-in fluids in situations where the well formation close to the reservoir is susceptible to acid attack. The base (alkaline) ingredient is magnesium oxide. It is added to a conventional (e.g., calcium chloride, sodium chloride, sodium bromide) brine in amounts of up to 5%.

Barito Basin

In South Kalimantan, Indonesia, a potentially abundant source of **CBM**. BP, ExxonMobil and **Pertamina** all have an interest. Barito is the largest, in terms of reserves, of the six possible sites of CBM production in Kalimantan: close behind is Kutei. The balance of conventional and unconventional gas in Indonesia is relevant to two major facilities: the **Bontang LNG plant** and the proposed Kalimantan–Java pipeline. Since 2005 there has been interest in such a pipeline but also doubts about its soundness in financial terms, and the state of affairs at the time of writing is that CBM production in Kalimantan 'looks likely to revive long delayed plans for a Kalimantan-Java pipeline'[8]. (*See also* **Jatibarang Field**; **Sanga-Sanga block**.)

Barracuda Field

Offshore Brazil, the scene of operation of an **FPSO** with **turret mooring**. It is in 835 m of water. It uses polyester mooring lines. The vessel can hold 0.34 million barrels and can receive up to 34 flexible risers. At the exploration stage the **pay** was found to be 153 m.

Barrhead, Alberta, CBM at

Longford Energy (HQ in Calgary) produces **CBM** from a reserve in Barrhead, Alberta. Initial drilling was to a **TVD** of 1145 m. A horizontal sidetrack from this well, of length 1289 m, was drilled and a CBM reservoir of very good permeability accessed.

Barton Field

Oil field offshore eastern Malaysia. The water depth is 40 m and the field is becoming mature. There are 20 production wells, one of which has been treated by a **swellable elastomer** as part of a stimulation project which also involves **water injection**. Reservoir pressure had dropped from 1000 p.s.i. to 500 p.s.i., with consequent loss of production.

Basin centred gas

A term equivalent to tight gas. (*See* **Chinook Ridge**.)

Bauxite, use of in hydraulic fracture

Bauxite particles can make an effective **proppant** for hydraulic fracturing. Pre-treatment at high temperatures is necessary to impart the hardness required for such use. Ultraprop®, manufactured by Saint-Gobain in Arkansas, is an example. Its median particle diameter is 0.74 mm.

Bazergan Field

Oil field in Iraq. Like so many oil fields in Iraq, Bazergan is undergoing rehabilitation and the US oil field services company Weatherford are active in this regard at Bazergan as well as at the Rumaila Field as noted in another entry. Twenty new wells are being drilled over a two to three year period that began in the second half of 2010. The new wells are each expected to produce twenty thousand to thirty thousand barrels per day. Other necessary work includes repair to oil and gas pipes having entered a state of neglect, to which the war and sub-standard maintenance before the war both contributed. Iraq is currently producing about 2.3 million barrels per day of crude oil and needs international help in order to raise that to a target of 6 million barrels per day by about 2015. (*See* **Majnoon Field**.)

BCD

Below Chart Datum. An indication of depth in shipping operations, including dredging, and also of the height of a tide. There is some variability in the position of the 'datum' but the level of the sea at low tide is a common one. Anything below the sea has a negative BCD. (*See* **J.F.J. de Nul**.)

Belokamenka

Near Murmansk in northern Russia, and the site of a major **FSO** through which 15 million tonness of crude oil - about 100 million barrels - can pass in a single year. It receives oil from the **Prirazlomnoye Field**.

Ben Avon

Jack-up rig currently operated in Nigerian waters by Canadian Natural Resources and a competitive rig. It can drill in water depths of up to 75 m and to well depths of up to 6000 m. Its **drawworks** are rated at 2000 h.p. and its **derrick** can take a load of 450 tonnes as can that of **GSF Adriatic VI** which has also seen recent service in Nigerian waters. (*See* **Olowi Field**.)

Bengkulu Basin

In Sumatra, seen as a source of **CBM** contributing to the potentially huge availability of that commodity in Indonesia. Coal from the basin has a calorific value, according to reports in its sales information, of 5500 kcal per kg equivalent to:

$$5\,500\,000 \text{ cal kg}^{-1} \times 4.2 \text{ J cal}^{-1} \times 10^{-6} \text{ MJ J}^{-1} = 23.1 \text{ MJ kg}^{-1}$$

The calorific value of methane is 889 kJ mol^{-1}, or:

$$(889 \text{ kJ mol}^{-1}/0.016 \text{ kg mol}^{-1}) \times 10^{-3} \text{ MJ kJ}^{-1} = 55.6 \text{ MJ kg}^{-1}$$

The intrinsic value of CBM is clear from this comparison. The reader should be aware that in addition to the CBM in Indonesia described in various entries in this dictionary there is CBM awating production at **Sulawesi** in a quantity of about 55 billion cubic metres. An obvious possibility is that any associated gas from the oil wells being drilled offshore Sulawesi could, after **condensate stripping**, be admitted to the same stream as the CBM. Neither oil nor CBM production at Sulawesi is at this time sufficiently advanced for the feasibility of this to be assessed.

Benguela Belize

Compliant tower oil production facility offshore Angola. It is in 390 m of water, a depth which would not have precluded a **fixed platform**. The *raison d'etre* of a compliant tower is its ability to flex, in contrast to a fixed platform, which is rigid. A compliant tower can only withstand a limited number of movements and adjustments before fatigue becomes evident. With the Benguela Belize more than 80% of the 'fatigue life' was taken up by installation, which involves moving the tower with a crane and lowering it into the sea with all that that involves in terms of forces on the structure. The facility can therefore only operate for the remaining 20% of the fatigue time. There will also be a compliant tower at Lombito Tomboco, an oil field offshore Angola, being developed by Chevron. Water depths there are between 3600 m and 6000 m. (*See* **Solan**.)

Bentley Field

In the UK sector of the North Sea, to the west of Shetland. The water depth is 113 m. Although it was discovered in 1977 it is only now being prepared for production, with a number of oil field development companies involved. BP will take the oil after production for marketing. Several exploratory wells were drilled at Bentley over the period 2003 to 2007, and **spudding in** of an appraisal well, which was eventually drilled to 1251 m, took place in

December 2007. An accessible **pay** of 15 m was encountered. The formation is sandstone, of good **porosity** and permeability, and this augurs well for production. There will be a conventional tanker moored at the scene which will receive oil from the production wells via a specially built **jack-up** rig. Transfer from there to shore will be by **shuttle tanker**. There will be wells for **water injection**.

Bentonite

Mineral substance very widely used in water based drilling muds. The dominant constituent is montmorillonite, the composition of which is expressed by the formula:

$$(Na,Ca)(Al,Mg)_6(Si_4O_{10})_3(OH)_6nH_2O$$

For drilling mud use bentonite will be milled down to a particle size of below 1 mm, typically 100 mesh (0.15 mm). The drilling mud density required depends on the well pressure, and muds for high pressure use containing bentonite will be in the 15 **ppg** to 20 ppg range. A **clay yield** of up to about 25 $m^3\ t^{-1}$ is expected with bentonite, higher than values achieved with **attapulgite**.

Benzoic acid

Simple organic compound which, as well as being used in bridging agents, as noted, is also used in water based **hydraulic fracturing fluids**. After such use it decomposes into benzene, readily soluble in crude oil, and carbon dioxide, which easily escapes, so it does not leave a residue.

Beryl Field, hydraulic fracture at

In 1989, the first hydraulic fracture by Mobil in the North Sea was carried out at the Beryl Field. The formation to which hydraulic fracture was applied had a permeability of 10 millidarcy (md) and the measured depth (MD) of the hydraulic fracture was 4900 m. The **PI** was raised from 0.7 barrels per day per p.s.i. to 1.4 barrels per day per p.s.i. (*See also* **Hydraulic fracturing fluid**; **Southern North Sea, tight gas in**)

Bettis-DD

A method of desiccating natural gas by means of a zeolite molecular sieve in pellet configuration. Like **MOLSIV™** it also acts on contaminants other than water. One data pair from the information on the manufacturer's web site will be examined. At a total pressure 1015 p.s.i and at 40°F (4°C, 277 K) the moisture content after desiccation will be 9 lb per million cubic feet.

Number of moles of methane in 1 million cubic feet at 277 K, 1015 p.s.i. pressure = $\{(1015/14.7) \times 10^5\ \text{N m}^{-2} \times 10^6\ \text{m}^3 \times 0.028\ \text{ft}^3\ \text{m}^{-3} / (8.314\ \text{J K}^{-1}\ \text{mol}^{-1} \times 277\ \text{K})\}$
$= 8 \times 10^7$ mol

Now at 277 K the saturated vapour pressure of water is 0.007575 bar

Weight of water in the methane in the absence of the desiccant =

$8 \times 10^7\ \text{mol} \times (0.007575 \times 14.7/1015) \times 0.018\ \text{kg mol}^{-1} \times 2.205\ \text{lb kg}^{-1}$
$= 348$ lb

So removal is 97%

(*See also* **Silica gel, use of in natural gas dehydration.**)

Bezinal®

Bekaert Zinc Aluminium, an alloy resistant to corrosion. It is 95% zinc, so to coat a wire for use in the sea in platform support with Bezinal® is equivalent to galvanising it. Bezinal® is in fact so used.

BHA

Bottom hole assembly. (*See also* **Drill collar**; **Downhole motor**; **Drilling jar**; **Reaming**; **WOB**.)

Bicone

Self-explanatory term for a drill bit having two cones. Introduced as long ago as 1909, the bicone configuration is now less common than the tricone (three cones). Tricones having been taken out of service through wear are sometimes reconditioned for resale. (*See* **MX**.)

Bi-Di

Range of **utility pigs** from Weatherford International[9]. A Bi-Di **pig** has a number of components and, as a unit, can turn a bend of **centre line radius** 1.5 × the outer diameter of the pipe, which represents a tight bend. This is made possible by the modular structure of the pig, different parts being capable of independent movement when a bend is encountered. Such a set-up is sometimes called a 'pig train'. The cleaning surface is made of polyurethane. (*See* **Super Dry 2000/2000S**.)

Big Foot Field

Oil and gas field in the Gulf of Mexico, under development with a view to beginning production in 2013. The water depth is 1500 m. Exploratory drilling was from the **Cajun Express** and an appraisal well was drilled by the **semi-submersible rig** Ensco 7500. The well from which production is expected to commence is a sidetrack of **TVD** 7654 m. It is intended that production will be at a **tension leg platform** which will also have drilling facilities. The drilling facilities are in obvious anticipation of expansion, as is the pipeline from the production unit, which when initially laid will have a capacity of 100 000 barrels per day.

Bilbao Knutsen

LNG tanker having entered service in 2004. (*See* **Pioneer Knutsen**.)

BIOCIDE RX-1225

Agent for killing of **SRB**, from Roemex (HQ in Aberdeen) and used at the seawater injection stage: it is therefore 'prevention' rather than 'cure'. It can be used on a batch or continuous basis. The latter would involve levels in the region 50p.p.m.to 100 p.p.m. Also from Roemex comes RX-270, which is **gluteraldehyde** based and controls SRB and **APB**.

Bio-Clear™ 242D

A Weatherford pipeline care product, capabilities of which include the killing of **SRB**. It contains glutaraldehyde and a quaternary ammonium salt. Its stablemate Bio-Clear™ 250 is also effective in killing SRB and its composition is 50% **glutaradehyde**. It is used in water at a level of 0.01% by volume.

Bir Seba

Oil field in Algeria (an OPEC country), where **spudding in** of the first appraisal was in late 2005. The eventual depth of the well was just under 4000 m. PetroVietnam Exploration and Production Corporation, PTT Exploration and Production (Thailand) and Sonatrach (UK) have been participants in the development since then and production is expected to commence in the final quarter of 2011 at 20 000 barrels per day, increasing to 36 000 barrels per day by 2014.

Bit Booster®

Device under development by a consortium of companies including Varel International for use with a **PDC** bit. The Bit Booster® will provide a force on to the bit in partial substitution for the **WOB**. The ultimate in performance of such a device would result in **drilling at zero WOB** and this the developers see as being realisable. Meanwhile the value of the device is that it will enable **directional drilling** to be carried out at low WOB. Movement of the bit is by a **downhole motor**.

BKB

Below kelly bushing. In well drilling power is provided to the 'rotary table' which is connected via a bushing to a device called the kelly (note the lower case) which in turn is connected to the drill pipe. When on initial drilling the depth of the well reaches that of the drill pipe – which will traditionally be 30 feet – the kelly is raised and another piece of drill pipe of the same length is attached between it and the original piece of pipe by means of a screw thread. This is called 'making a connection' and has to be repeated every time a length equivalent to that of a single section of drill pipe has been drilled. A natural datum point for expressing the well depth is therefore the level of the kelly bushing and so the acronym BKB is widely used. Re-use of drill pipes is possible but requires a close inspection of their condition because of the obvious consequences of failure. Since the 1980s the procedure using a rotary table and kelly has been increasingly replaced by top drive drilling. Elimination of the kelly means that the length of a piece of drill tube which can be put in place in one connection is limited not by its dimensions but by that of the **derrick**, typically reducing the number of connection operations by a factor of three. (*See also* **Saipem 10000**; **Zhana Makat Field**.)

Black powder

Term applied to fine particles arising from corrosion of the interior surface of a gas pipeline. Such particles are obviously easily entrained in the gas and can lead to blockages further downstream. The treatment for black powder is application of a gel, for example one from the **SureGL™** range.

Blackford Dolphin

Semi-submersible rig having recently been in service in the **Jubilee Field**. It can operate in water depths up to 2000 m and can drill to depths of 9000 m. The **derrick** can take a load of 680 tonnes and its **drawworks** are rated at 4920 h.p.

Blackpool, Lancashire

For generations a budget holiday resort and previously a fishing town, Blackpool is in 2011 the scene of exploration for **shale natural gas**. At a site a few miles from Blackpool, there has been exploration which has involved hydraulic fracture. If the enterprise is successful it will be the first production of shale natural gas in the UK.

Blake Field

In the UK sector of the North Sea, producing oil and gas since 2001. The water depth is 95 m and the operator is Talisman. Oil from the field is taken to an **FPSO** which also receives oil from the Ross Field. The field is being extended, and an important factor is the high wax appearance temperature (WAT) of the oil. This has been addressed by insulation of the pipe taking the oil to the FPSO: once it is there the high WAT can be addressed by use of an additive which works analogously to a simple freezing point depressant.

Blane Field

Straddling the UK and Norwegian sectors of the North Sea, producing oil and associated gas since 2007. The water depth is 70 m and the operator is Talisman. The two production wells are tied back to Ula, a **fixed platform**, operated by BP. Current production at the Blane Field is about 12 000 barrels per day. An eventual target of 17 000 barrels per day is aimed for. The **Maersk Guardian** has seen service there.

Blenheim Field

Oil field in the North Sea about 230 km from Aberdeen, discovered in 1990 and believed to contain 53 million barrels of crude oil, making it a relatively small field. Production (commenced 1995) is by an **FPSO** in 148 m of water with **shuttle tanker** transfer of the oil. The operator is Talisman.

Blowout preventer (BOP)

Often a so-called BOP will be a blowout preventer stack, that is, an assembly of blowout preventers in a single installation. It might contain as many as six **ram-type blowout preventers** and a smaller number of **annular blowout preventers**, the latter above the former. In the specification of drilling rigs the rating of a blowout preventer can be represented as XK, where X = the pressure rating in thousands of p.s.i. So a 15K blowout preventer would have a rating of 15 000 p.s.i. (*See* **Offshore Freedom**, **West Epsilon**).

Blue Angel

Initially a supply vessel only, having recently been converted to a **well stimulation vessel** and bearing the Norwegian flag. It has **dynamic positioning** and carries methanol in a quantity of up to 135 tonnes. Some of the other details of its functions are yet to be confirmed.

Blue Dolphin

Said to be the largest well stimulation vessel currently in service and based in the Gulf of Mexico. It can also carry 48 000 litres of acid (requiring dilution before use) and 1250 tonnes of proppant. A well stimulation vessel carries the ingredients of some the fluids it needs for blending on board, and on Blue Dolphin the blending unit can provide up to 23 000 h.p.

Bluell formation ND

A carbonate formation currently being developed for oil production under not the most favourable of circumstances: thin **pay** zones and a permeability less than 2 md. A vertical hole was drilled to a depth of 305 m with horizontal drilling through a pay as thin as 1.5 m, there being barefoot **completion**. The low permeability per se was not seen as necessitating any form of acid treatment but drilling mud deposition had reduced the permeability even further, and so along the horizontally drilled portion the **matrix acidising** agent SXETM, a Schlumberger product composed of 70% hydrochloric acid and 30% diesel with an emulsifier, was applied in conjunction with MaxCO^3 Acid™. MaxCO^3 Acid™ is also from Schlumberger, and is for carbonate formations only. This is all 'work-in-progress' and figures for production from the Bluell formation ND are not yet available. We conclude this entry by noting that the term 'hydraulic fracture' has been applied in company literature to other procedures involving acid degradation of a carbonate formation. However, no **proppant** was used at the Bluell formation ND, where action was purely chemical. Moreover, in a description of the simple acid stimulation at the **Polecat Creek well TX** the term 'frac fluid' was in one coverage used to describe the acid reagent. Obviously, the present author is not an arbiter of correctness in terminology. He is, however, hopeful that readers of this dictionary will be alert to ambiguities[10].

BMR™

Loss of productivity of an oil well might be due not to depletion or pressure loss but to the poor condition of the well resulting from deposition of solids, for example, from **drilling fluids**. Such an effect will restrict flow, to the impairment of well performance. A well showing evidence of such restriction can be 'rehabilitated', and there are many chemical products for this. **BMR™** from CETCO Drilling Products is one such. It removes **bentonite** and, as a bonus, natural clay that has found its way into a gravel pack. It is used at concentrations of about 0.33 **ppg** in water. (*See* **Nu-Well® 220.)**

Boil-off

Term applied to the loss of LNG by natural evaporation in storage and handling. Such losses are significant and have to be recovered. A boil-off gas compressor will withdraw the gas whilst it is still below its critical temperature, that is, when it is distant from phase equilibrium with the liquid at its normal boiling point, and apply pressure to return it to the liquid phase. Any LNG terminal will be set up to reverse the effects of boil-off. Compression will not, of course, return the gas to LNG as the temperature is well above the critical temperature of methane. The **Mizushima LNG**

import terminal has two boil-off gas compressors, each with a capacity of 8.1 tonnes per hour, in heat terms:

$$[(8.1 \times 10^3 \text{ kg}/0.016 \text{ kg mol}^{-1}) \times 889 \text{ kJ mol}^{-1}/3600 \text{ s}] \times 10^{-3} \text{ MW} = 125 \text{ MW}$$

That is the rate at which heat would be released if the gas was burnt at the same rate as it was being produced at the compressor. For fuel supply rates to be so expressed – production and burning in a steady state – is quite standard whether or not that is the actual arrangement. The **North Adriatic LNG Terminal** also has two such units. (*See also* **Dabhol LNG terminal**; **Map Ta Phut LNG Regasification Terminal**.)

Bolivar Coastal Field

At the eastern edge of **Lake Maracaibo**, Venezuela, and producing for over eighty years. It now has about 7000 wells and current production is around 3 million barrels per day of oil and 0.41 million cubic metres of gas. Note that Lake Maracaibo is significantly saline having between 5000 mg and 15 000 mg of sodium chloride per litre of water depending on the season. (*See* **LL652 Field**.)

Bombay High North Field, glycol dehydration at

It has been reported as a case study that the **GDU**s at Bombay High North Field began to require something like four times the usual quantity of glycol in order to function. 'Usual' at these particular GDUs was about 0.12 gallons of **TEG** per million cubic feet (20 litres of TEG per million cubic metres) of gas desiccated. Follow-up included analysis of the glycol after use and review of the process whereby, after use, the glycol and water are separated so that the former can be re-used. The case study is incomplete at the time of writing this entry, and factors possibly responsible for the glycol loss which is undeniably occurring are being considered in turn. These include foaming of the TEG.

Bonaparte gas pipeline

In Australia's Northern Territory, of 287 km length. It is wholly within the Northern Territory, not crossing any boundaries. The pipe supplier gives the delivery as 30 PJ per annum – petajoules per year, where peta denotes 10^{15} – of gas. This is reviewed in the simple calculation in the box below.

$$[30 \times 10^{15}/365] \text{ J day}^{-1}/37 \times 10^6 \text{ J m}^{-3} = 2 \text{ million cubic metres per day}$$

The Northern Territory has only one major centre of population and many scattered smaller ones. This is reflected in its energy requirements. The pipe sections comprising the pipeline were made in Queensland. (*See also* **Roma to Brisbane pipeline**; **Sunrise Field**.)

Bond™Lite

Resin coated ceramic **proppant** from Carbo™Ceramics in Houston. In general with proppants, a resin coating prevents formation of fines of the ceramic which jeopardise the conductivity, and this is so with Bond™Lite. The conductivity is 7023 md-ft at a closure stress of 2000 p.s.i. and 1432 md-ft at 14 000 p.s.i. (*See also* **Fracture pressure**; **Prime Plus™**; **SandWedge®**; **Vernon Field**.)

Bonga

Oil field offshore Nigeria, in which Shell have an active presence. It was discovered in 1993 and has been producing since 2005 by means of an **FPSO** which is permanently moored. The water depth is approximately 1000 m. The FPSO, which can hold up to two million barrels of crude oil, uses steel catenary risers (SCR). These are more commonly associated with the **tension leg platform** or the **spar platform** (e.g. **Devil's Tower**) and use of the SCR at the FPSO at Bonga was a first. There are also facilities for **water injection**. The hull of the FPSO was constructed in Korea and sailed to Tyneside, England for building of the topside. (*See also* **Erha Field**; **Nitrate ions, action on by SRB**.)

Bonny Island LNG plant, Nigeria

Commencing operations in 1999 and now having six LNG trains. Data for the plant when all six trains are operating are 22 million tonnes of LNG per year and 4 million tonnes of LPG per year, requiring 3.5 billion cubic feet per day of gas intake. A mass balance is attempted below.

Molar mass of the input gas = $(0.056 \times 4/26) + (0.016 \times 22/26)$ kg = 0.022 kg, where the molar mass of LPG has been taken to be that of butane.

Volume of input gas per day = $[(26 \times 10^9\ \text{kg}/(0.022\ \text{kg mol}^{-1} \times 40\ \text{mol m}^{-3})]/365\ \text{m}^3$
= 0.081 billion cubic metres or 2.90 billion cubic feet.

The difference of 0.6 billion cubic feet per day (17%) is of course due to the condensate of higher boiling range than LPG.That major amounts of such condensate *are* produced at Bonny island is documented, and there are storage and offloading facilities for it. The mass balance above can be seen as being quite precise. A seventh train is expected to come into operation at Bonny Island in 2012, which will raise the total production capacity of the plant to about 30 million tonnes of LNG per year.

Bontang LNG plant

In East Kalimantan, Indonesia, receiving gas from a number of fields. It commenced operations as long ago as 1977 and has over that time provided 200 million tonnes of LNG for Japan, Korea and Taiwan. It has eight LNG trains, the most recent of which came into service in 1999. Having exceeded the **Arun LNG plant** in production thereby becoming the largest LNG plant in Indonesia Bontang encountered difficulties of its own, partly due to uncertainties in the supply of gas through under-performance at the fields on which it draws. Even so, a ninth train is planned. LPG from Bontang is exported to Japan. The 2005 figure for production from Bontang was 20 million tonnes of LNG. More recent years have seen this decline. (*See also* **Barito Basin**; **Sanga-Sanga block**.)

Boomvang

Oil and gas field in the Gulf of Mexico (water depth 1124 m), having a **spar platform** of the same name which, instead of a cylindrical hull, has a truss arrangement making the platform a **truss spar** which, like any spar platform, has its base well above the seabed. It moored is by wire lines. The Boomvang truss spar has an 'identical twin' at the adjacent Nansen Field, and these two truss spars were the first such to come into service. The production capability of each is 40 000 barrels per day of oil and 5.6 million cubic metres per day of gas.

Borgland Dolphin

Semi-submersible rig constructed in Northern Ireland, and a 'rebuild'. It can operate in up to 455 m of water and can drill to a depth of 8500 m. The chain by which it is moored was developed expressly for Borgland Dolphin. The rig has spent a good deal of its time in the Norwegian sector of the North Sea.

Borgsten Dolphin

An early **semi-submersible rig** still in service. It can operate at water depths up to 300 m and has **spread mooring**.

Borrox®

Range of **reaming** tools from Smith Bits, a Schlumberger company, available with a **tungsten carbide** insert or a diamond cutting surface according to the hardness of the formation. The term 'fishing' usually means retrieving a blocked drill bit the presence of which is preventing further drilling. The term does however have a related but different meaning in oil well engineering and this entry provides a basis for explaining that. A tool such as a reamer will itself need 'retrieving' after use and will accordingly have a 'fishing neck' which attaches to the body of the tool to allow such retrieval. The Borrox® reaming tools come in a range of sizes. Those intended for use in holes between 5⅞ inches and 6¼ inches (determined of course by the size of drill bit previously used) have a fishing neck diameter of 4¾ inches. Those intended for use in holes in the range 12 inches to 17½ inches have a fishing neck diameter of 10 inches. These figures have been taken from the two extremes of the manufacturer's specifications and there are intermediate bore size/fishing neck diameter data. In length terms, the smallest model has a tool length of 77 inches and a fishing neck length of 25 inches. The top-of-the-range has a tool length of 112 inches and a fishing neck length of 49 inches.

Botswana, oil products in

A landlocked country lacking oil reserves is likely to be dependent on pipelined refined products. This is true of Botswana, which receives all of its petroleum material by pipeline from South Africa. Of course, the most likely origin of those pipelined products is oil imported by South Africa, so in effect South Africa is 'exporting' refining services to Botswana. Time will tell whether the abundance of oil in Angola, now an OPEC country, will make any difference to countries such as Botswana. Obviously, for this to happen Botswana would need to get into the downstream industry. Lesotho, also landlocked, obtains imported refined products from South Africa.

Boulton Field

'Marginal' gas field in the UK sector of the North Sea, operated by Conoco-Phillips, and off the Lincolnshire coast. Production began in 1997. There are two production platforms, each of which is a **Seaharvester**. **Directional drilling** was used for the production wells. These had **TVDs** up to 4000 m and **AHD** values of 3050 m. The reader is referred to two other entries – that for AHD and that for **Tortuosity, in well drilling** – for background on the calculation apropos the Boulton Field below.

At the Boulton Field the angle Θ is given by $\tan^{-1}(3050/4000) = 37°$

The measured depth (MD) is $4000/\cos 37° = 5010$ m.

Taking the tortuosity (τ) to be 1 degree per hundred metres of the MD, a value of 50.1° is obtained. The **DDI** is given by:

$$DDI = \log[(MD \times AHD/TVD)\tau] = 5.3$$

which can be compared for example with the range 5.56 to 5.91 for three wells off California given in one of the entries referred to and the value of 8.279 for the BD-04A well offshore Qatar

The term *step-out* is sometimes used as a synonym for AHD, as in the source from which the above depth figures for the Boulton Field came. The appreciable values for the calculated DDI for the Boulton Field are connected with the lengths and depths involved, each of which is several kilometres. (*See* **Njord Field**.)

Brady®

Proppant and gravel pack material having the same chemical composition – 99.48% silicon dioxide – but particle sizes that depend on the application. In either **proppant** or gravel pack use the conductivity will be a factor, the higher the better. (*See also* **Fracture pressure**; **Vernon Field**.)

Brage Field

In the Norwegian sector of the North Sea, sea depth 140 m. It has a **fixed platform** which is set up for drilling as well as for production. It was discovered in 1980 and is now at a stage where methods for improved recovery are being applied. (*See* **Sognefjord**.)

Braided line

A multi-strand alternative to **slickline** for getting tools in and out of wells. Because of the braiding it has some voidage and its weight per unit length is lower than that of a slickline of the same materials and diameter. (*See* **Eclipse Wireline**.)

Brent Charlie

GBS in the Brent Field in the North Sea, in a water depth of 141 m. It has been in operation since 1976. In January 2010 it received repair to the deck by SPS Overlay. (*See* **Conkouati**.)

Bridgeport CT

As yet, vehicular LNG usage in the US is mainly limited to 'projects' that will presumably also be seen as pilot schemes in preparation for wider usage. One such project is centred on a refuelling station in Bridgeport which, along with conventional products, will supply LNG as a fuel for a fleet of eighteen vehicles used in refuse collection. The LNG comes from a local supplier that is a subsidiary of GDF Suez.

BRINEDRIL-N®

Drilling-in fluid from Halliburton having a brine base and dispersed polymer. A supplementary product is BARACARB®, which is a bridging agent made from marble. From the same 'stable' comes BARADRIL-MAG™, also for drilling-in, containing magnesium oxide as a bridging agent. (See **SAFE-CARB**.)

Bruce Field[11]

In the North Sea, producing oil, gas and condensate. It is operated by BP and by BHP Billiton. There are 11 production wells and sufficient reserves to justify tripling that number or more. Current practice at the field is to reinject gas after **condensate stripping**. Production for 2005 was 5 million barrels of oil and 4 billion cubic metres of gas. (*See* **Keith Field**.)

Bruna

A.k.a. Fiume Bruna, the first **CBM** well in Italy, having undergone **spudding in** in 2010. The coal structure, of 1 to 2 millidarcy permeability, was shown to be suitable for hydraulic fracture and **proppant** admittance. The **pay** of CBM is at a depth of 340 m. There is also **shale natural gas** and this predominates at greater depths. The co-existence of the two types of unconventional gas - CBM and shale natural gas - makes statements of gas yield per ton more difficult to interpret than for the CBM fields in Virginia discussed in the entry for **Virginia, CBM in**, for which a value of up to 600 cubic feet per ton was given. That being understood, the value of 152 cubic feet per ton for the field in which Bruna has been drilled can be seen as being below the values for the Virginia fields which, are amongst the most abundant methane yielders of any coal fields in the world. A value of 152 cubic feet per ton does not therefore signify non-viability. (*See also* **Fort Yukon AK, CBM at**; **Virginia, CBM in**.)

BT1,2,3,3H

Series of tricone bits from Bit Brokers International. Operating speeds are 50 r.p.m. to 200 r.p.m. for BT1, 50 r.p.m. to 150 r.p.m. for BT2 and 40 r.p.m. to 100 r.p.m. for BT3 and BT3H. The respective **IADC** codes are 121, 211, 311 and 321. A first digit of 1, 2 or 3 in the code denotes a drill bit made from steel but not to the exclusion of steel bits *faced* with **tungsten carbide** which is in fact the case with BT3. (The second digit in the IADC code depends upon the Moh hardness and the third on the bearing.) (*See* **Matrix body**.)

Bualuang Field

Offshore Thailand, operated 100% by Salamander (HQ in London). The two exploratory wells Bualuang 05 and Bualuang 01 had a **TVD** of respectively

1220 m and 1270 m. The 05 well has a gross **pay** of 25 m and a net pay of 21 m, the 01 well 23 m gross and 20 net. The water depth is 60 m and the **FPSO** 'Rubicon Vantage' is used in production. This can store 0.57 million barrels of oil. As a result of the low pressure at the Bualuang Field the oil is not only free of dissolved methane but has also lost some of its own low-boiling constituents. Oil in this state is called 'dead oil' and it also occurs in a minor degree at the **Douglas Field** off the UK. (*See also* **Laos, oil exploration in**; **Sinphuhorm gas field**)

Buchan Field

Oil field in the North Sea, water depth 111 m, that has been producing since 1981. The **porosity** is 9% and the permeability 38 millidarcy. The **pay** of oil is 585 m. The production facility is a converted **semi-submersible** drilling rig (the 'Drillmaster') having undergone major structural upgrading. By 1983 the field was producing 32 000 barrels of oil per day. In 1996, having become much less productive, it was acquired from BP by Talisman. An **FPSO** was taken to the field and **coiled tubing drilling** applied in an attempt to obtain previously inaccessible oil. It was the first time coiled tubing drilling had been carried out from an FPSO base. Production became 13 000 barrels per day and further coiled tubing drilling is expected to raise this, enabling the field to be productive at least until 2018. Depletion over the 30 year period of production is such that **depletion drive** is carried out at the field with gas which, before production began, was a component of the liquid phase in the reservoir.

Buckland Field

Oil field in the North Sea, producing since 1999. Its year of 'peak production' was 2000, when it yielded 28 000 barrels per day. There is also associated gas. Initial proposals to use an **FPSO** at the Buckland Field were not followed up, and instead oil from the production wells is sent by means of a bundle of pipelines to a platform at the Beryl Field. **Water injection** takes place at the Buckland Field, but in a rather unorthodox way. The reservoir water has an ionic content such that scaling in the formation would occur if it was used in water injection. Scaling of this sort can jeopardise permeability and offset the stimulating effects of water injection. At the Buckland Field therefore water free of such ions from an aquifer at the nearby Beryl Field is transferred to Buckland by an **electric submersible pump**.

Buffalo Venture

FPSO in long-term service at the field of the same name (Buffalo) in the Timor Sea where it began operations in 2000. A conversion from an oil tanker originally built in 1976, it has single point mooring and a **POB** of 19.

Bukhoosh Field

Oilield offshore the UAE, discovered in 1969 and producing since 1974. It is currently operated by Total. It is close to the **Umm Shaif Field** and linked to it by a pipeline. It supplies gas to the **ADGAS Plant**.

Burghley Field

In the North Sea, close to the **Auk Field** and being developed jointly with it. The water depth is 143 m and the owner of the field is Talisman. Two new wells are scheduled to be drilled at Burghley, oil from which will be taken to the **FPU** Balmoral. Oil from the Auk Field will be taken to infrastructure at the **Fulmar Field**. (*See* **Glamis field**).

Butterfly valve

Used in oil pipelines, such a device in its simplest form is a quarter-turn valve comprising a plate which occupies the entire cross section of a pipe thus stopping flow altogether. One 'quarter turn' enables unrestricted flow to be resumed. A ball valve is also a quarter turn valve, fluid passing through the hollow part of the spherical valve. One quarter turn isolates the hollow part so that fluid encounters the blank side and cannot progress. Clearly, whether in the open or closed position, the ball requires supports ('seats'). These can be made of PTFE for ball valve applications where temperatures do not exceed 200 °C. The 'seat' in a butterfly valve is around the edge of the plate and seals the closure by an 'interference fit' between the plate circumference and the inside wall of the pipe. There are a number of elastomers which have been used to make seats for butterfly valves. Butterfly valves and ball valves find extensive use in oil pipelines and the ball valves also have an important place in well engineering. (*See also* **RING-O®**; **Utility pig**.)

Byford Dolphin

Semi-submersible rig, built in Norway in 1979. It has **spread mooring** and can operate at depths of up to 1000 m. It is currently operated by BP under contract. (*See* **Solan**.)

Byford Dolphin rig (courtesy: Svein Åge Berge)

C

Cajun Express

Semi-submersible rig in the Transocean fleet. It was built in Singapore in 2000. Capable of operating in water depths up to 2600 m, it has **dynamic positioning** but is also configured for conventional mooring using wire rope when operating in shallower waters. It can drill subsea to a depth in excess of 10 500 m and has a **POB** of 130. (*See* **Big Foot Field**.)

Cal Canal

Onshore gas field in California. The Chevron 88-31 well at the field has a depth of 3.83 km, **pay** 541 m. Cal Canal has been productive since 1978.

Calcium, dominance of in oil well cement

In oil well cement, calcium is present as silicates, aluminate and aluminoferrite. These are all ionic compounds and constitute the clinker that comes from the cement kiln. Blending with gypsum – more calcium – precedes milling to a suitable particle size. One cement can usually be distinguished from another by microscopic examination, and features so observed correlated with such properties as curing time. The curing involves hydration of the inorganics in the cement with added water.

Camisea Gas Field

In Peru, discovered by Shell in 1986, and commencing production in 2004. The field is undergoing expansion. Its gas is distributed long distances by pipeline for domestic and industrial use, and also to copper mines in southern Peru using a specially installed pipeline. It provides gas for the **Peru LNG Project**.

Cancelled LNG terminals

The table below gives a (not exhaustive) list of planned LNG terminals in the US that failed ever to get off the ground.

Location	Details
Brownsville TX	Cheniere Energy had entered into an agreement with the Port Authority of Brownsville for a terminal. No further development. Cheniere *do* have a terminal at nearby **Sabine Pass**.
Near Portland, ME	The regasification facility was to have been on a former US Navy site.
Hope Island, ME	'Local opposition'
Vallejo, CA	Operation of this would have involved the transfer of LNG tankers, with Coast Guard escort, under the Golden Gate Bridge.
Near Mobile, AL	Intervention by the Governor of Alabama and withdrawal by ExxonMobil.
Near Eureka CA	'Local opposition'. The regasified product was to have been used to make electricity at 200 MW.

The matter of siting of LNG terminals, or other facilities where large amounts of hydrocarbon will be present, will always involve push-pull between oil companies, governments and community groups. In respect of LNG terminals two points can be made. One is that the record of LNG shipping over a period of fifty years is unblemished: there is no known case of death or injury during LNG shipment since such shipment began in 1959 with a consignment from the Gulf Coast to the Thames Estuary. Secondly, at about the time of 9/11 concerns were expressed that LNG tankers could become targets for terrorists. That might have been a factor in failure of the Vallejo facility to become a reality. (*See* **Kitimat LNG terminal**.)

Cantarell Field

In the Gulf of Mexico, the largest offshore oil field in the world. Production began in 1980 and in 2009 was 0.77 million barrels per day, well down on the figure of 2.1 million barrels per day that was being realised in 2004. There is associated gas. The operator is Pemex Exploración y Producción. The field owed its early productivity (until about 2000) to a natural gas bubble that maintained the reservoir pressure, often seen as having been unique. More recently the same function has been fulfilled by nitrogen injection at 33.6 billion cubic metres (39 million tonnes) per day. The nitrogen is produced at an onshore facility and admitted at a pressure of 115 bar to a pipeline of diameter 36 inches for conveyance to a production platform 53 miles offshore. It then enters injection wells with a depth of about 1400 m. Some of the associated gas is reinjected. Well stimulation activity is planned so as to raise the production above a million barrels per day within the next two to three years. The Jay Field in Florida – very much smaller than the Cantarell Field – has for 25 years used nitrogen injection *and* **water injection**. (*See also* **Casing exit system**, **Chicontepec Field**; **Hawkins Field**; **Nitrogen rejection**.)

Capixaba

FPSO, made by conversion of a previously existing unit at Keppel shipyard in Singapore. In fact it was previously a passenger vessel in the Stena fleet and based in Norway. The Capixaba can produce up to 100 000 barrels per day of crude oil and can operate at water depths up to 1350 m. Its **water injection** capacity is 138 000 barrels per day. This figure is briefly examined below.

> In the entry on water injection *per se* it was pointed out that in planned oil production in Iraq it is expected that use of 12 million barrels per day of water will increase the oil production rate by 9.5 million barrels per day giving a ratio:
>
> $$9.5/12 = 0.79$$
>
> The corresponding ratio for Capixaba is 100000/138000 = 0.72
>
> The agreement is reasonable and supports the view that 20% to 30% of water so used does not reach the oil at which it is targeted.

Capixaba is currently operating in the Cachalote Field offshore Brazil. Its production will extend to the 100 000 barrels per day given above as the number of wells connected to it increases. When in operation at full specification it will be connected to six production wells and three injection wells. It has facilities for gas dehydration using **TEG**.

Captain Field

Oil field in the North Sea, operated by Chevron-Texaco. The water depth is 104 m and it produces 85 000 barrels per day with an **FPSO** (also called 'Captain') as the production facility. The **gas cap** contains about 400 million cubic metres of natural gas. Captain Field is the scene of the longest horizontal gravel pack – 2234 m – ever installed in a well. That was in 2001. There are facilities for **Glycol dehydration** by use of **TEG**.

Carbon dioxide flooding

Carbon dioxide occurs at high purities in carbon dioxide fields in several parts of the USA, and can be admitted to oil wells to enhance recovery. The current price is $2 to $3 per thousand cubic feet. An example is the Bravo Dome formation in NM, believed to contain 224 billion cubic metres of carbon dioxide of purity greater than 99.9%. Carbon dioxide from Bravo Dome is taken by pipeline across the border to Texas for use at some of the oil fields there in enhancing recovery. The pipeline summed as its NM and TX parts is 210 miles long and it has one **compressor station**.

Carbon dioxide storage in oil fields, comments on

Carbon dioxde removed from post-combustion gases, for example at coal fired power stations, requires storage, and disused oil fields have been suggested as storage sites. There is also a 'movement' to inject carbon dioxide into oil fields to increase the pressure and hence the production. The view of the author is that these two points need to be considered separately, and we begin with the second. As is clear from so many entries in this dictionary, it is straightforward to reinject associated gas in order to maintain pressure, so why 'import' carbon dioxide? Carbon dioxide injection takes place in the **Lula Field** at the Brazilian government's insistence. This is believed to have *raised* the production costs of hydrocarbon from the field not least because of the need to separate associated gas and carbon dioxide and reinject the latter. At an offshore field it is also straightforward to inject seawater. A similar operation at an onshore field might be possible if there is an aquifer. So the assertion that carbon dioxide sequestration at producing oil fields would *benefit the oil industry* is not a convincing one. As for storage in disused oil fields, the difficulty is that this might preclude the restoration to production of the field. As has been made clear elsewhere in this volume, when a field declines in production that is no proof at all of depletion. It might be that the wells are in poor shape and that workover will improve matters. More probably, it could be that the drill pipe structure is predominantly vertical and that horizontal drilling will access **pay** not previously within range. Again, there are numerous examples in this volume of fields so restored to good production levels. If such a field has previously been used to store carbon dioxide its rehabilitation as an oil producer is obviously less straightforward. **Carbon dioxide flooding** is a process distinct from sequestration.

Carbon steel, use of in well casing

The norm for pipe used as well casing material, though not to the total exclusion of other materials. Whatever the material, the casing is required to be seamless. Carbon steels differ in mechanical properties according to heat treatment received during manufacture. Different grades of carbon steel are used in the fabrication of well casing having, for example, different values of **yield strength**.

Cardiff (NZ) development

New Zealand will be requiring enormous amounts of energy for reconstruction following the 2011 earthquake, and that domestic reserves will suffice is inconceivable. It is, however, doubtful whether the NZ government will put on hold energy projects begun prior to the earthquake (and in any case there are contracts in place with developers which will not be negated by the earthquakes: at Cardiff the developer is TAG Energy), one of which was the Cardiff gas/condensate discovery in Taranaki, off the west coast of the North Island. Taranaki, on- and offshore, is already the NZ region of most productive of hydrocarbons. Gas from Cardiff can share existing pipeline facilities with established fields. Field development will involve horizontal drilling and hydraulic fracture as the gas is in the 'tight' category. (*See also* **Raroa**; **Southern North Sea, tight gas in**.)

Carter-Knox Field

Oil field in Oklahoma also very productive of gas and condensate, recently the scene of use of an **impregnated drill bit** in addition to other types of bit in the creation of deep wells. In fact, impregnated bits were used over the depth interval 4570 m to 5777 m. Production at the Carter-Knox Field began in 1923 and over that time it has yielded 27.6 million barrels of oil as well as the gas and condensate.

Casablanca Field

Offshore Spain in the Gulf of Valencia, discovered in 1983 and a major part of Spain's oil productivity. Recently a flexible pipeline manufactured by Technip has been installed to link two wells at the field. Having a length of 11 km, most of it will be 'flowline' positioned on the seabed, but parts will act as a **jumper**. The point is made in another entry that an advantage of flexible as opposed to rigid pipeline is that continuity between flowline and links to the production facility is possible. (*See* **Conkouati**.)

Casing exit system

Device for **window** installation in a well casing as the first step in sidetracking. The device will also have a role in **whipstock** positioning. The cutting device is called a *milling assembly* and might contain **PDC**. The **Cantarell Field** was the scene of a recent casing exit operation. The undertaking was extension of a well casing to access another reservoir and the milling assembly was the QuickCut™ downhole tool from Weatherford. This did the entire 'window opening' without renewal over a 7 hour period. Another recent success story for QuickCut™ was at the Lower Cotton Valley Field in Texas where it was used to create an exit at 3650 m and, again, fulfilled the task without renewal. The well casing material was **P 110**. (*See* **Prime Plus™**.)

Casing for oil wells

As drilling takes its course in a well, tubular casing is put in place. It provides support for further drilling and prevents collapse of the enclosure having been drilled. At intervals in the casing there will be O-ring seals. The casing can be cemented into place by admitting wet cement which on exiting the bottom of the casing will rise and form a sheath around the outer surface. Plain **carbon steel** is a common choice of material for casing as discussed (*see* carbon steel). The pH of the cement when wet is probably not a factor in cement performance *per se*, but a cement which is too acid can attack the O-ring seals. Movement of the casing during cementing will clearly make for both displacement of the casing from its intended positioning and unevenness in the cement sheath formed. During cementing therefore the casing is held in position by devices called centralisers. The number of centralisers used will vary from one exploratory operation to another. Having fulfilled its role in drilling the casing held by cement obviously has an important role in support once the well is complete and production of oil begins.

Caspian pipeline consortium

This conveys crude oil from **Tengiz** (Kazakhstan) to Novorossiysk (Russia), a distance of 1510 km, having since initial operation been extended from Tengiz to **Atyrau**. Ownership is 51:49 Rosneft and Shell. Total 2009 passage was 50 million barrels of oil plus 0.2 million barrels of condensate. At Novorossiysk, a port town, there is a terminal to which the oil is taken for transfer to tankers. (*See also* **Kazakhstan-China oil pipeline**; **Karachaganak Field**.)

Catcher prospect

In the North Sea. An initial appraisal well and a sidetrack from it found major **pay** of oil. A second sidetrack is planned. Significant amounts of associated gas are present. If and when production begins there might be a strong case for reinjection of this to sustain reservoir pressure.

Cauvery-Palar basin

Offshore southern India, only recently the focus of hydrocarbon exploration by Reliance Industries (HQ in Mumbai) with very promising results. At a water depth of 1194 m and a drilled depth of 3815 m, major **pay** of gas and condensate were discovered early in 2011. Exploration and appraisal by Reliance continue.

CBM

Coal bed methane.

(*See also* **Airth; Barrhead, Alberta, CBM at; Bengkulu Basin; Cheyenne Plains pipeline; Curtis Island; Fort Yukon AK, CBM at; Gladstone LNG project; Horseshoe Canyon; Huabei Field; Jatibarang Field; MER-14X; New South Wales, proposals for CBM in; Raton Basin; San Jaun Basin NM; Sanga-Sanga block; Sekayu; Taldinskaya CBM field; Ukraine, unconventional gas in; USA, selected CBM reserves in; Virginia, CBM in; Wales, exploration for CBM in**.)

CBM Drill-In™

Drilling fluid additive for **CBM** from ARC Fluid Technologies (HQ in TX). Its action depends upon imbibition of the substance by the coal, and consequently it is organic in nature. It has a flash point as low as 25 °C (77 °F), about the value one would expect from the naphtha fraction on the refining of crude oil. The usual formulation is 3 gallons of the additive per 1000 gallons of the **drilling fluid**, or 2000 p.p.m. (*See* **Fort Yukon AK, CBM at**.)

Ceduna Basin

Offshore North West Australia, the scene of proposed drilling at water depths up to 4600 m. BP are seeking licences to drill in the hope of commencing in 2013–2014.

Ceiba Field

Field offshore west Africa, the first in that region (in 2002) to be the scene of a **Frac-Pack** operation. This was at one particular well at the field ('Ceiba No. 5 well') only. The field is in 799 m of water and the **pay** is 100 m. The well to which Frac-Pack was applied was of **TVD** 2430 m and a total of 91 tonnes of **proppant** was used at concentrations up to 8 **ppg**. Frac-Pack was carried out using the Halliburton **well stimulation vessel** War Admiral. Current production at the field is about 12 500 barrels of crude oil per day.

Cendor Field

Offshore Malaysia, water depth 70 m. The field is currently the scene of operation of a **MOPU**. This is connected to an **FSO** with **spread mooring**. A further phase of development of the field is being planned, which is likely to lead to replacement of the existing arrangements with an FPSO. In general a MOPU is most likely to find application at a marginal field or at a large field where there is concurrent production and expansion.

Centerfire™

Modular **MWD** tool from General Electric having resistivity and γ-ray functions. It was recently used to good effect in drilling at the **Jidong Field**. In one well at the field it was used over the drilling depth interval 2120 m to 2759 m and enabled operators accurately to ascertain the reservoir depth. In another well at Jidong it was used over the depth interval 1630 m to 2125 m for the same purpose.

Central Asia–China gas pipeline

This carries gas from Turkmenistan through Uzbekistan and Kazakhstan to China. It is a dual pipeline (A and B) each of 1833 km length (1146 miles) and is fabricated of 42-inch diameter sections with a wall thickness of ⅛ inch. Having its origin just on the Turkmenistan side of the border of that country with Uzbekistan, it 'collects' gas from Kazakhstan fields including the **Karachaganak Field**, the **Kashagan Field** and **Tengiz** along its route. The breakdown of the length is:

Within Turkmenistan: 188 km

From the border between Turkmenistan and Uzbekistan to the border between Uzbekistan and Kazakhstan: 530 km

From the border between Uzbekistan and Kazakhstan to the border between Kazakhstan and China: 1115 km

Total 1833 km

The aimed-for annual rate of natural gas supply to China along the pipeline in 2013 is 40 billion cubic metres. By that time it will also be receiving, via a branch pipeline, gas from the **Aral Sea**. (*See also* **Kazakhstan-China oil pipeline**; **Myanmar-China pipeline project**.)

Centre line radius

When a pipe is bent, the axis of the pipe in the area of the bend forms the arc of a hypothetical circle. The radius of the circle is the centre line radius of the bend. In oil and gas pipelines the ratio of the centre line radius to the pipe outer diameter is the index of the degree of bending. Clearly the smaller this radius the sharper the bend: in the limit of a perfectly straight pipe it is infinity. (*See* **Bi-Di**.)

Cepu Field

Onshore field in Java, Indonesia, discovered in 2001 and producing since 2009. The formation at the field is carbonate, with **porosities** in the range 20% to 30% and permeabilities up to 200 millidarcy. Oil from the field is taken by pipeline to an **FSO** in the Java Sea. ExxonMobil and **Pertamina** each have a 45% stake in the field, the balance being in the hands of bodies including the local Javanese administration. Hydraulic fracture will feature in development of the field to a target production of 165 000 barrels per day by 2014.

Cetus

SSDV used in oil and gas fields. It has **dynamic positioning** and can carry up to 1500 tonnes of rock. It has recently been in service in Singapore waters, preparing the sea-floor for pipe lay.

Challis Venture

FPSO built in Japan in 1989, the first to use **SARLAM**. It has been the sole production facility at a field offshore Indonesia and has produced 40 million barrels of oil. The field (also called Challis) is now becoming depleted, so the future of Challis Venture is uncertain. At the time of going to press it is believed to be on its way to Singapore to await its fate there.

Champion Field

Offshore Brunei, water depth 30 m. The operator is Shell, and current production is 50 000 barrels per day. The field contains 500 reservoirs[12] and has a correspondingly complex production structure with 30 platforms and 185 subsea pipelines. Some of the wells are drilled in a 'weaving' pattern so as to access several narrowly separated reservoirs. The production figure is modest when viewed against the huge infrastructure and this raises a question in the mind of the oil analyst as to viability. Viability may be enhanced by the fact that exploration well drilling is taking place at the field continually. Newly discovered oil could, one would imagine, be tied back to existing facilities and increase the return on them. (See **Rasau Field**.)

Changbei shale natural gas field

Tight gas field in China, producing since 2007 with Shell and PetroChina as the operators. Well drilling was very challenging, being required to include horizontal and multi-lateral parts. The gas supplies centres of population within China, including Beijing.

Chayvo field

Oil field offshore Sakhalin, the first in the huge Sakhalin development to become productive. At Sakhalin conditions are challenging; temperatures can descend to minus 40 °C. Oil from wells owing their existence to **extended reach drilling** is brought ashore to a land facility called Yastreb, permanently installed on Chayvo beach. In addition to its production role it is home to those working at the site when they are on duty, having accommodation quarters. Production began in October 2005 at 50 000 barrels per day and this had risen by a factor of five by February 2007. There are also large amounts of associated gas. The operator is Exxon Neftegas and the **ExxonMobil Fast-Drill Process** was used at Chayvo. The reader will appreciate that this approach is very helpful in extended reach drilling primarily for the **ROP** it provides for and also because of the reduced likelihood of bit failure with consequent loss of time through replacement. (*See also* **J.F.J. de Nul**; **Orlan** and **TVD**.)

Cheyenne Plains pipeline

This takes gas from Wyoming to Kansas, from the Rockies to the Midwest. Like many onshore gas pipelines in the US it uses pipe of diameter 36 inches. There are two **compressor stations** with a combined rating of 22 164 h.p., and the pipeline typically conveys about 25 million cubic metres of gas per day. In the Rocky Mountain states generally coal bed methane (**CBM**) is not always clearly distinguished from natural gas and the payload of the Cheyenne Plains pipeline will undoubtedly include some CBM. (*See also* **Curtis Island**; **X80 steel**.)

Chicontepec Field

Onshore oil field in Mexico, production at which is modest (thirty to forty thousand barrels per day) but which might offset decline at the **Cantarell Field**. The operator is PEMEX. **Pay** is 350 to 400 m. Reservoir **porosity** and permeability are low, a factor making for difficulty with expansion of production. The field has multiple reservoirs, at depths typically of 3000 m. Where oil is, within a single field, distributed between many reservoirs, that necessitates a large number of wells. That reserves at Chicontepec are huge is not in question, and Permex has the support of the Mexican government in setting the extraordinarily high target of 1 million barrels per day by 2014, a thirty-fold increase over a three year period. Because of these factors, this will require the drilling of 20 000 wells. (*See* **Champion Field**.)

Chikyu

Drill ship built by Mitsubishi and able to work where the sea depth is up to 2500 m. At such depths **dynamic positioning** is indispensable and this is provided by the same thrusters that are used in propulsion. Its drill works are supported by a **derrick** of height 70 m. Its **drawworks** are rated at 5000 h.p. It was delivered by Mitsubishi to its owners in 2005, and having seen service offshore Australia will in 2011 be used to drill offshore Sri Lanka.

China, acquisition of FPSOs by

China is developing offshore oil fields and over the period 2010 to 2013 is expected to invest $US23 billion in **FPSO** units some of which will be new and some modifications and rebuilds of existing units.

Chinook Ridge

Straddling the Alberta and BC boundary, the scene of **basin centred gas** production by Shell. Peak production (attained only over a short period) was 0.5 million cubic metres per day. Expansion is viable but at the present time, and although restricted by limited infrastructure, Shell is considering amoungst other proposals that there should be an LNG train.

Chuchupa

Gas field offshore Colombia, operated by Chevron. Production there began in 1979 and by 2009 cumulative production was 70 billion cubic metres. There are two production platforms which are also set up for drilling. These are in 60 m of water. Three new horizontal wells were drilled by Chevron at Chuchupa in 2006, and there are now fifteen production wells in all: seven horizontal, seven vertical and one directionally drilled. Gas from these, which is free of condensate, goes to the two production platforms. The reservoir depth is about 1700 m and the wells vary in production by an order of magnitude, from 0.2 million cubic metres per day for the vertical wells having been drilled early in the field's production life to 2.5 million cubic metres per day for the most recently drilled horizontal wells.

Clay yield

Volume of drilling mud (in m^3) obtained from 1 kg of dry mineral constituent. Typical values for muds with **bentonite** or with **attapulgite** as the major constituent are given in previous entries. An organic additive such as **ODC-15™** might be used to enhance the clay yield. The term 'mud yield' is an arguably preferable but less widely used synonym.

CleanStim™

Hydraulic fracturing fluid from Halliburton undergoing trials at oil fields in several parts of the US including Texas. The trials are concerned largely with performance over a temperature range having regard to temperature rises experienced by fluids descending wells.

ClearPAC

Gravel pack carrying fluid from Schlumberger. When a gravel pack is installed, any residue from the medium that carried the gravel impairs the effectiveness of the pack by lowering its permeability. A helpful performance criterion for the carrier fluid is the ratio of permeability (usual in units of millidarcy) of a pack having been laid by carrying fluid to that of one not having been exposed to any carrying fluid, other things being equal. With ClearPAC this ratio exceeds usually exceeds 0.9 although the 'other things' that have to be 'equal', including the nature of the gravel itself and its particle size range, are numerous. Like its competitor product **Hydropac℠**, ClearPAC contains a gelling agent. (*See also* **Perdido platform**; **South Furious Field**.)

CMO

'Cubic Mile of Oil', approximately what the world uses per year[13]. This can be tested against the more conventional figure of 80 million barrels per day as shown in the shaded area below.

$$80 \text{ million barrels per day} \times 365 \text{ days per year}$$

$$\downarrow$$

$$(80 \times 10^6 \times 365 \times 0.159)/10^9 \text{ km}^3 \text{ per year}$$

$$= [(80 \times 10^6 \times 365 \times 0.159)/10^9\,] \times (5/8)^3 \text{ cubic miles per year}$$

$$= 1.13 \text{ cubic miles, equivalent to a cube of side } (1.13)^{1/3} \text{ miles} = 1.04 \text{ miles}$$

so the 'CMO per year' rule of thumb holds up. A hypothetical strategic petroleum reserve holding a year's supply of crude in a spherical tank would require a tank radius:

$$[(3/4\pi) \times 1.13]^{1/3} \text{ miles} = 0.65 \text{ miles } (1033 \text{ m})$$

These figures of course are for crude oil. On fractionation, refinery gain to a degree of something like 7% would apply.

Coiled tubing drilling

Procedure whereby a drill bit is connected to a hydraulic motor receiving **drilling fluid** down a hollow tube. The hollow tube is admitted to the well as the penetration depth increases from a reel containing possibly thousands of metres. As an example, the Canadian company Technicoil (formed 1997) manufactures drilling coil ranging in diameter from ½ an inch to 3½ inches. It is made from ribbon steel – steel plate heated to form a roll – which is then unravelled and made into a continuous tube before again being heated and coiled around a support ('drum'). Technicoil make available coiled tube in this form in lengths of up to 7000 m. In addition to making coiled tubing for drilling available to the market, Technicoil has six rigs of its own for coiled tube drilling, each carried by a heavy duty truck when being moved between drilling sites. In its own drilling the company uses 2⅞ inch tube in lengths, when unravelled, of up to 1800 m. Coiled tube drilling is usually to depths which by the standards of *rotary* drilling are moderate: Technicoil's deepest 'conquests' in the USA and Canada respectively are 2281 m and 1980 m. (Note that CDT – the acronym for coiled tubing drilling – is a trademark of Schlumberger.)

(*See also* **Buchan Field**; **NK-12c**; **QT-800**; **Sidetracking, by coiled tubing drilling**; **Slimhole drilling**.)

Collet-Grip™

Connector for in situ sections of subsea pipeline from HydroTech, available in a wide range of sizes. It unites two sections of pipe by a Viton® seal mounted in a firmly bolted support. This, where the device is being used in repair, necessitates removal of any coating on the existing pipes. It differs from the split sleeve clamp in that there is no split, and bolting parallel to the axis[14] compresses the body of the device at the position of the Viton® seal.

Columbus Field

Not as one might have thought in Ohio (where J.D. Rockefeller first set up in the oil business) but in the UK sector of the North Sea. It is a condensate field, that is, a non-associated gas field rich in condensate (a.k.a. natural gas liquids, NGL). The **pay** is 40 feet (12.2 m) and is at a vertical depth of about 3000 m from the sea-floor. Columbus is in the **ETAP** and tieback to infrastructure there is planned, for example, condensate after separation will be taken to the **fixed platform** at the **Lomond Field** for storage prior to being taken ashore. The Lomond Field connects with the **Forties pipeline**, one of the most heavily used anywhere. (*See* **Maule Field**.)

Completion

This term when applied to oil and gas wells means the choice and installation of the structure whereby hydrocarbon from the reservoir enters a well. The simplest arrangement, simply called *open hole completion* or *barefoot completion*, is for the well casing to continue up to – not *in*to – the reservoir. Where this approach is taken, the rock structure of the well has to be sufficiently strong for self support. If this is not so, *conventional perforated completion* is the method used. With this the casing is continued into the reservoir and has a role in supporting it, hydrocarbon entry being by means of holes in the casing pipe wall. *Sand exclusion completion* uses a **gravel pack screen** to filter out sand where it would otherwise be carried along with the hydrocarbon. With oil wells, *drain hole completion* is an option. It is achieved by having a horizontal or slanting passage between the reservoir and the well casing. From the brief description of 'completion' in this entry of the dictionary it will be clear that the site of entry into the reservoir can easily be damaged. The role of a completion fluid is to protect it. Drilling muds are not suitable for use as completion fluids, partly because the solids they contain can, on deposition, do harm. One expects a completion fluid to be homogeneous and the most basic ones are simply brines. The same is true of a drill-in fluid, use of which precedes that of a completion fluid. Factors relevant to choice of completion fluid or drill-in fluid include viscosity and pH. The latter is important where the rock formation is easily attacked by acids, in which case a substance such as **BARADRIL-MAG™** can be incorporated into the brine base fluid. Viscosity adjustment can be obtained by means of products such as **SAFE-VIS™**. Loss of drill-in fluid or completion fluid to openings in the formation can be prevented by use of a bridging agent. A well once 'completed' can be held in reserve until circumstances at the reservoir are clearer, as at the **Prirazlomnoye Field**. (*See also* **Arcasolve™**; **BARADRIL-MAG™**; **Bluell formation ND**; **Captain Field**; **Frac-Pack**; **Fracture pressure**; **GPT**; **Perforating gun**; **SAFE-CARB**; **Slimlne drilling**; **Triethanol amine** (endnote)).

Compressor station

When natural gas is being conveyed long distances by pipeline, compression along the route will be necessary to sustain the flow at a suitable rate. For this it passes through a compressor station (a.k.a. a compression station). The compressor will be driven in one of three ways: with a gas turbine (as at **Werne, Germany**), with an internal combustion engine (as at **Davenport Downs**) or by electricity. In the first two cases, gas from the pipeline will be used as fuel. When natural gas is being conveyed in the US network a compressor station will be encountered about every 50 to 100 miles. (*See also* **Carbon dioxide flooding**; **Golden Pass LNG terminal**; **Pumping station**; **Rio de Janeiro-Belo Horizonte Gas Pipeline II**, **South-East to North-East pipeline**; **Texas Eastern Transmission**.)

Condensate, density of

In the range 50 degrees to 120 degrees API[15], or 780 kg m^{-3}down to 560 kg m^{-3}. (*See also* **Aceh**; **Angel Field**; **Jebel Ali Refinery**; **Yttergryta Field**.)

Condensate stripping

Natural gas condensate consists of hydrocarbons in the approximate range C_2 to C_{12}. The heavier compounds are separable from the gas by cooling at atmospheric pressure. The lighter ones are, even under the cooler conditions, well above their critical temperatures, so compression at that stage will bring them into the liquid phase. Ethane (critical temperature 32°C) might if present in major amounts be separated from the other components. Indeed, where **stabilised condensate** is the desired product such separation will be necessary. Worldwide, 3.6 million barrels of natural gas condensate per day are produced. Although the carbon number range in condensate is not as wide as that in crude oil, condensate clearly can be fractionated. At Ras Laffan in Qatar 146 000 barrels per stream day (bsd) of natural gas condensate are being so processed. The lightest constituents yield 8000 bsd of LPG, and 63 000 bsd of naphtha, and the heavier ones 52 000 bsd of kerosene suitable (in cloud point terms) for use as jet fuel and 24 000 bsd of diesel. An obvious advantage of condensate refining is that there is no heavy residue. The kerosene is exported and the diesel retained for the domestic market. Note that production of LPG from condensate is standard practice at many gas fields even if the heavier components are not separated from each other: this is the intention at the **PNG LNG project**. At the time of writing, condensate is attracting prices in the range $US60–70 per barrel: crude oil is a little over $US100 per barrel. A gas that has undergone **condensate stripping** is called 'dry gas'. If condensate is not so stripped but enters a pipeline with the gas it might require pigging for removal, as at the **Nam Con Son Field**. (*See also* **Aceh**; **Angel Field**; **Bonny Island LNG plant, Nigeria**; **Kollsnes**; **Bontang LNG plant**; **Bruce Field**, **Golden Pass LNG terminal**; **Hides Field**; **Jebel Ali Refinery**; **Kårstø processing plant**; **Stabilised condensate**; **Tangghuh LNG plant**; **Trent Field**; **Yttergryta Field**.)

Conkouati

FPSO currently in service in the Yombo Field offshore the Republic of Congo. It was converted in 1991 from a tanker, so the hull had done a great deal of work and much repair and restoration work was necessary, which might by traditional means have been uneconomic. The hull was in fact renovated by a process called SPS[16] Overlay. Introduced in 2000 and having the endorsement of Lloyd's Register, this uses a composite of two metal plates bonded together with an elastomer and involves no 'hot work'. Repairs to Conkouati by this means took place without taking the vessel out of service at all. Many other FPSOs and also **FSO**s have been repaired by this method including the FPSO in service at **Zaafarana Field**. Not only vessels but platforms can also be repaired by this means, as with the **Brent Charlie**. (*See also* **GSF Grand Banks**; **Sedneth 701**.)

Connate water

Water trapped in the pores of a geological formation which, once oil or gas is released from the pores and phase equilibrium is lost, itself exits the pores and becomes one component of produced water. Gas exiting the pores will carry water vapour which on condensation becomes an additional component of produced water. Water of neither category is expected to condense at reservoir temperature but will do so during well passage. (*See* **Watered out well**.)

Constitution/Ticonderoga Field

Oil field with associated gas in the Gulf of Mexico, at a water depth of 1500 m. The reservoir permeability is a high 800 md, and the **porosity** is also high at 28%. Production is at a **spar platform**. Initial exploratory drilling was at Constitution only, and when about 135 m of **pay** were found at the nearby Ticonderoga Field the original design specifications for the spar platform were changed so that Ticonderoga could be tied back to it. Like all such the spar platform does not itself rest on the seabed but having settled above it by natural buoyancy is then attached by moorings. (*See* **Angot Field**.)

Cook Inlet

In Alaska, producing oil and gas over a very long period. The point at which vessels enter the Inlet is 68 miles from the state capital Anchorage. Cook Inlet was the site of the first **monopod platform**. This came into operation in 1967[17] and, simply called *Monopod*, is one of sixteen platforms to have been installed at Cook Inlet. Others include *Tyonek*, on the north side of the Inlet, which is an abundant producer of gas. Bruce platform, further into the Inlet, produces oil and gas. Produced water is returned to the saltwater of the Inlet, and some of the associated gas is flared. Decline does seem to be the keynote of affairs at Cook Inlet at present. In addition to the closure of the **Kenai LNG plant**, Chevron is withdrawing its involvement at Cook Inlet and offering its assets there for sale. Chevron's recent production at Cook Inlet has been modest enough: 4000 barrels of oil per day and 2.5 million cubic metres of natural gas per day.

Corrib Field

Gas field offshore the Republic of Ireland operated by Shell. At the time of going to press **Solitaire** is in service there, laying a pipeline to a gas terminal near Glengad. A related calculation follows.

It has been stated in previous entries that a **pipe laying vessel** will lay 4 to 5 miles per day of pipe. It is not however 'laying pipe' continually. The vessel, which will usually have **dynamic positioning**, will need re-positioning as pipe laying takes its course. Welding of one section of pipe to another might not be fast enough to keep passage through the stinger continuous. The pipe once installed has to withstand high pressures which will have been 'designed into' the pipe specification, but that the design stress of the virgin metal will apply at a weld cannot be assumed. This necessitates close examination of welds and possible further welding. The time in days required for a pipe laying job cannot therefore be obtained by dividing the number of miles of pipe by some number between 4 and 5.

It is reported that at Corrib Solitaire will be in action 24/7 for six weeks. The pipeline which it will install will be 83 km (52 miles). The time in laying, during which pipe is passing through the stinger, is approximately:

(52 miles/4.5 miles per day) = 12 days to the nearest day.

The percentage of the time spent on pipe admittance to the sea is then:

$(12/42) \times 100\ \% = 30\%$ approx.

Cossack Pioneer

FPSO having **turret mooring**. Its operator is Woodside Petroleum and it has over a long period received oil and gas from the Australian North West Shelf, operating in a sea depth of 80 m. There are ten production wells. Natural gas produced by Cossack Pioneer goes to a liquefaction facility at Karratha in NW Australia. LNG produced there goes to markets in Japan and South Korea. The fields of the North West Shelf have 'over-performed', and Cossack Pioneer will not in the longer term be adequate. Its replacement will be a 'rebuild' FPSO with a **disconnectable turret**. The rebuild will be in Singapore of a vessel previously at Sakhalin. (*See also* **J.F.J. de Nul**; **PTT PCL**.)

Costa Azul LNG terminal

On the Baja California coast, commencing operations in 2008. All of the US LNG terminals are on the east side of the continental landmass, making Costa Azul the only such facility on the west side (although of course it is not a US facility, as Baja California belongs to Mexico). In June 2010 Costa Azul was the scene of the first delivery of LNG from the **Peru LNG project.** (*See also* **Altimira LNG terminal**, **Cove Point, MD**; **Kitimat LNG terminal**.)

Cottonwood Field

In the Gulf of Mexico 138 miles from the coast of Texas, sea depth 670 m, operated by Petrobras. A point of engineering interest is that it is a **long-distance subsea tieback** to a 'host platform'. Gas from two wells at the Cottonwood Field is taken 17 miles along a 6 inch diameter pipeline to a **fixed platform** operated by W&T Offshore. There it becomes 'third-party gas' and is taken to shore by pipeline after **condensate stripping**. Amounts of condensate from the Cottonwood Field are major. (*See* **Durango well**.)

Cove Point, MD

Scene of a receiving terminal for LNG. Mothballed for over 20 years, it now has an output of more than 30 million cubic metres of regasified product per day. The LNG terminal at Elba, GA was also mothballed for about 20 years, and since re-entering service produces about 12 million cubic metres of gas per day. The LNG receiving facility at Everett, MA has been in continuous service since 1971, never having been mothballed. Cove Point, Elba and Everett all currently receive gas from Trinidad and Tobago although they all long predate natural gas production in T&T. The three facilities in this entry are all of course on the eastern side. 'Planned and proposed' LNG terminals include some for the east coast, such as that at **Weaver's Cove MA**, but also for the Pacific coast, as imports from Malaysia, Indonesia and Australia become more important. These will build on the capacity provided by the **Costa Azul LNG terminal**. There are many LNG terminals for the Pacific coast and for the Gulf coast with 'planned', 'approved' or 'pending' status at the present time and there are formidable hurdles to be crossed before these can be included in the LNG reception capacity of the US rather than join the list of **cancelled LNG terminals**. (*See* **Kitimat LNG terminal**.)

Crude oil, stabilisation of

In order to avoid explosion hazards in shipping and handling, crude oil might have to be denuded of its lowest boiling components. It must be remembered that the material which becomes LPG in refining is still present in crude oil exiting a well. The Reid Vapour Pressure below which such hazards are eliminated is something like 10 p.s.i. Stabilisation can be by stimulation of evaporative loss of the lighter components by evacuation of the space above the liquid. This can be performed in steps until the target RVP is attained. (*See also* **Abqaiq, crude oil stabilisation at**; **Forties pipeline**; **Kashagan Field**; **Stabilised condensate**.)

CrudeSep®

Device for cleansing **PFW** before its return to the sea, used at a number of installations in the North Sea. It incorporates a number of devices including a cyclone, a centrifuge and a skimmer. It is capable of reducing the oil content of PFW to 10 p.p.m. or less. Its use has recently extended to north-west Australia, where in custom built form it is used for PFW removal at the Stag and Apache platforms. The 10 p.p.m. target for PFW having been cleansed by the unit was in trials achieved without difficulty. Regulations at the north-west shelf require that treated PFW when discharged into the sea shall not exceed 50 mg oil per litre of water at any time and shall average less than 30 mg of oil per litre over a 24 hour period. Neglecting differences in density between oil and water, milligrams per litre and parts per million weight basis are numerically equivalent, so that the CrudeSep® is performing adequately is clear. (*See* **Stena Spey**.)

Culzean discovery

In the UK sector of the North Sea in 92 m of water. Maersk drilled an appraisal well in 2008 and confirmed the existence of gas and condensate at a depth of 4500 m. The original appraisal well was then sidetracked to obtain information on possible **pay** to the east of the well. **Gryphon**, later to encounter some difficulty as recorded, was used in the appraisal well drilling.

Curtis Island

In Queensland, Australia, the scene of proposed installation of multiple (up to eight) LNG trains to convert coal seam gas (CSG a.k.a. **CBM**) to LNG. There is an obvious difference between CSG and natural gas: the former contains no condensate, and it is clear from several of the entries in this volume that condensate is a commercially very significant product the absence of which puts CSG at a disadvantage. If the eight trains become a reality the production will be 20 million tonnes per annum of LNG. There is already an agreement whereby Sinopec will take 4.3 million tonnes per year. Dredging at Port Curtis will be required to support the proposed LNG production. (*See also* **Horseshoe Canyon**; **Gladstone LNG project**; **MER-14X**; **New South Wales, proposals for CBM in**; **Sanga-Sanga block**.)

Cusiana Field

In Colombia, producing since 1994 with BP as the operator. Production facilities at commencement were capable of receiving up to 40 000 barrels per day. There are three stacked reservoirs and the formation is sandstone. Permeabilities are in the range hundreds to thousands of millidarcy. When discovered in 1992 Cusiana was believed to be a major find, a view which subsequent events did not support and one press coverage described the field at these early stages as being 'too small for Wall Street'! The theme there in 2011 is 'depletion management'. (*See also* **Meji Field**; **Tyra Field**.)

Cutter suction dredger

One of a number of types of dredger, finding application *inter alia* in the oil and gas industry. There is frequently a need to smooth the seabed in offshore production, for example to ensure that pipelines lie flat. Dredging is also used where vessel access to an onshore facility is needed, to widen or deepen channels to accommodate ships as at **Curtis Island**. A cutter suction dredger is stationary when on dredging duty, being held by one or more spud devices similar to those often used in the support of **jack-up** rigs. The cutting head can, when in position on the sea floor, be moved by a swinging action through wide arcs. Obviously the performance of a cutter suction dredger depends upon the seabed structure, in particular whether it is sedimentary or non-sedimentary. A cutting tool is selected according to the hardness of the seabed. The cutting function of the dredger is followed by its suction action, and material loosened by cutting is drawn into a suction pipe. Traditionally, the diameter of the suction pipe is used to classify the entire vessel. Such diameters are in the range 100 mm to 1500 mm, so one might talk of a 'thousand millimetre dredger'. Delivery from the suction pipe is into barges and sometimes the debris is used in reclamation. Half a million cubic metres of debris in a week's operation of a cutter suction dredger is not uncommon. Note that this will have a bulk density depending on particle size and the degree of seawater uptake. A *trailing suction hopper dredger* like the **Orisant**, in contrast to a self-propelled cutter suction dredger, has no mooring but draws layers of debris from the seabed by trailing suction pipes whence the debris is transferred to a hopper. (*See also* **J.F.J. de Nul**; **Map Ta Phut LNG Regasification Terminal**.)

CySep™

Separator for oil and gas, suitable for subsea use (though not restricted to such use) using cyclonic action. It can also remove seawater, making it in fact a three-phase separator. Water so separated can of course be reinjected into the well. If that is not the intention and the water is to be readmitted to the sea ('water discharge' in the language of the offshore industry), CySep™ can reduce its oil content to 40 p.p.m. A sister product is **G-Sep CCD™**, a two-phase cyclonic separator which separates oil and gas and also provides for scrubbing of the gas to remove such constituents as carbon dioxide and hydrogen sulphide. (*See also* **PFW**; **CrudeSep®**.)

Dabhol LNG terminal

One of three such facilities in India, receiving LNG from Qatar, Oman, the United Arab Emirates, Australia, Malaysia and Indonesia. Its output of gaseous fuel will be used in electricity generation. It is 180 km south of Mumbai. Having only just been commissioned the facility receives 1.2 million tonnes per annum of LNG. A simple calculation along the lines of those performed in relation to the **Sodegaura LNG terminal** follows below.

$$\text{Rate of electricity production} = 1.2 \times 10^{9}\text{kg} \times 55 \times 10^{6}\ \text{J kg}^{-1} \times 0.35/(3600 \times 365 \times 24)\ \text{W} = 700\ \text{MW}$$

Expansion of reception to 5 million tonnes per year is expected, and at that stage not all of the regasified product will be used in electricity generation. **Boil-off** at the Dabhol LNG terminal is recovered by two compressors. (*See also* **Fos Tonkin (Fos-Sur-Mer) LNG Terminal**; **Hazira LNG terminal**.)

Dagang Field

Onshore oil field in China, producing since 1994. Over the early period of production one of the three pipelines at the field was found to be corroding rapidly whilst the other two were not corroding to an unexpected degree. Amongst the factors considered was **SRB**. At an onshore field SRB if they are present must have come from from the reservoir. Water from the formation at which oil to the corroding pipeline originated was tested for sulphate content and found to be 50 mg/L to 150mg/L in SO_4^{2-}. The middle of the range becomes in p.p.m. terms:

$$(100 \times 10^{-3}\text{g}/96\ \text{g mol}^{-1})/(1000\ \text{g}/18\ \text{g mol}^{-1}) \times 10^{6}\ \text{p.p.m.} = 19\ \text{p.p.m. molar basis}$$

This is a lower value than expected in seawater but a long way above being a mere 'trace', indicating further scope for investigation of whether SRB was a factor in the corrosion observed.

Damietta

Egyptian port and the scene of construction of a facility for LNG production. It will use 'APCI Technology' and will initially have a single 'train'. (*See* **MCR®**.)

Dapeng LNG terminal

At Guangdong on the southern coast of China, having received its first shipment in 2006. It receives LNG from Australia and from Qatar. It can receive 3.7 million tonnes per annum of LNG, equivalent to a gas supply rate of 15 million cubic metres of 'regasified' product per day. This makes its capacity comparable to that of the **Rudong LNG terminal**. Sixty-five per cent of the gas from Dapeng is used in electricity generation. A calculation similar to that given in detail for the **Dabhol LNG terminal** reveals that the rate of power generated will be about:

$$[0.65 \times 15 \times 10^6\ \mathrm{m^3} \times 37 \times 10^6\ \mathrm{J\ m^{-3}}/(24 \times 3600\ \mathrm{s})] \times 0.35 \times 10^{-6}\ \mathrm{MW} = 1450\ \mathrm{MW}$$

The gas not used to make power is distributed to users in places including Hong Kong.

Davenport Downs

In Queensland Australia, the scene of a new gas **compressor station**. Compression is by two 126 kW (169 h.p.) engines. Its capacity is given as up to 175 TJ per day which converts to:

$(175 \times 10^{12}\ \mathrm{J\ d^{-1}}/37 \times 10^6\ \mathrm{J\ m^{-3}})/24\ \mathrm{h\ d^{-1}} = 200\ 000\ \mathrm{m^3\ h^{-1}}$ where the $\mathrm{m^3}$ is at 1 bar and 288 K. The equivalent in mass terms would be:

$$200\ 000\ \mathrm{m^3\ h^{-1}} \times 42\ \mathrm{mol\ m^{-3}} \times 0.016\ \mathrm{kg\ mol^{-1}} \times 10^{-3}\ \mathrm{t\ kg^{-1}} = 135\ \mathrm{t\ h^{-1}}$$

David Tinsley

Jack-up rig having been in service since 1981, built by Mitsui and part of the Noble fleet. It can operate in water depths up to 90 m and can drill to 7500 m. After a four-year period in Qatar it was returned to Sharjah in the UAE where **UWILD** preceded assessment for expansion of the accommodation facility to raise the **POB** to a maximum of 150.

DBNPA

2,2-dibromo-3-nitrilopropionamide, a general purpose biocide which has been used to good effect in **SRB** control. (*See* **Glutaraldehyde**.)

DDI

Drilling difficulty index. (*See also* **Boulton Field**; **Tortuosity, in well drilling**.)

DDS™

Drill string Dynamics Sensor. (*See* **PWD**.)

DEA

Diethanolamine. (*See* **Amine process**.)

Deborah Field

Gas field In the southern part of the UK sector of the North Sea. Recently (Q4 2010) approval was given for creation of a natural gas storage facility at the field which, it is hoped, will enter service in 2015. The storage facility, operated by Eni, will have a capacity of 4.6 billion cubic metres and its introduction will double the gas storage capability of the whole UK. The number of barrels of oil to which this amount of gas is equivalent in energy terms is[18]:

$$(4.6 \times 10^9 \text{ m}^3 \times 37 \times 10^6 \text{ J m}^{-3}/44 \times 10^6 \text{ J kg}^{-1}) \times (1/900 \text{ kg m}^{-3}) \times (1/0.159 \text{ m}^3 \text{ bbl}^{-1})$$
$$= 27 \text{ million barrels}$$

Clearly the magnitude is such that the facility would, if kept full, act as a strategic reserve analogous to those for oil, which by government authority could be drawn on in a contingency if there were interruption to supply from UK or Norwegian fields.

Deep Blue

Pipe laying vessel built by Hyundai in 2001 and owned by Technip Marine. Having **dynamic positioning**, it can operate in water depths up to 2500 m, can carry pipe at weights of the order of thousands of tons, and lays pipe according to the **J-Lay** pattern. (*See* **Reel laying of pipelines**.)

Deep Panuke Field

Offshore Nova Scotia, under development at the time of going to press. The **pay** of gas at the field is 69 m. Drilling and **completion** have involved the **jack-up** rig **Rowan Gorilla III** which, in its application at Deep Panuke, has had **spud can** support. Production will be at a unit having single buoy mooring.

DeepSTIM III

Well stimulation vessel operated in the North Sea by Schlumberger. Having **dynamic positioning**, it can hold a quantity of 770 tonnes of **proppant**. It can discharge liquid into a well at up to 50 barrels per minute

Deepwater Champion

Newly built (at the Hyundai shipyard in South Korea) drill ship, expected to make its debut in the Black Sea when ExxonMobil will be the operator. It can function in up to 3600 m of water and drill to subsea depths of up to 12 km. Its **derrick** can hold a load of 1800 tonnes. (*See also* **Deepwater Discovery**; **Skandi Aker**; **Tuapse trough**.)

Deepwater Discovery

Drill ship in the Transocean fleet, built in 2000. It can operate in water up to 3000 m and to well depths up to 9000 m. Its **moonpool** is 52.5 feet by 40.9 feet and it has **dynamic positioning**. Its **derrick** can take a load of 907 tonnes and its **drawworks** are rated at 5000 h.p. (*See* **Agbami Field**.)

Depletion drive

An oil reservoir initially containing no associated gas will start to release some previously dissolved gas as it becomes depleted. This can be directed at the movement of oil eliminating the need for other measures such as **water injection**, as at the **McClure Field**. It has to be distinguished from **gas cap** drive. Where the associated gas was initially present and its expansion as the oil depletes can be used to move the oil. (*See also* **Buchan Field**; **Sognefjord**.)

Derrick

Support structure for drilling. For onshore drilling there are a number of classifications, including the K type which applies to the ZJ30DBS in use at the **Zhana Makat Field**. The K type has three trusses – two vertical and one horizontal – forming a rigid structure. Offshore drilling facilities have equal need for a derrick, which will be a conspicuous part of a **semi-submersible rig** or a drill ship. The synonymous term 'mast' is more frequently applied to onshore rigs. (*See also* **Chikyu**; **Deepwater Champion**; **Drill ships, newly constructed**; **Discoverer Enterprise**; **DSS-20-CAS-M**; **GSF Celtic Sea**; **Kaskida Field**; **Leiv Eiriksson**; **Maersk Gallant**; **Maersk Guardian**; **Nini and Cecilie fields**; **Noble Amos Runner**; **Noble Leo Segerius**; **Noble Phoenix**; **Noble Roger Eason**; **Rowan Gorilla III**; **Saipem 10000**; **Skandi Aker**; **West Epsilon**; **West Navigator**.)

Desulphurisation of crude oil by SRB

In reservoirs, wells and pipelines **SRB** cause unwanted effects and have to be controlled. Once the oil is safely at a terminal, use of SRB is possible and there is some development interest in this on the Gulf Coast at present. Sulphur in crude oil is present as thiols, the sulphur analogue of alcohols:

$$R\text{-}SH$$

where R denotes an alkyl group and sulphur is in oxidation state +2. Its reduction by SRB could be summarised:

$$RSH + 2H^+ + 2e \rightarrow S + RH + H_2$$

sulphur having gone to oxidation state zero.

Devil's Tower

Oil field in the Gulf of Mexico where production is by a **spar platform** in 1710 m of water, with two steel catenary risers. The platform can receive from the risers up to 60 000 barrels per day of crude oil.

DEXTRID®

Filtrate control agent from Halliburton, based on starch and for use with water based drilling muds. The upper limit of temperature for its use is about 120 °C and it is incorporated at up to around 0.2% by weight, at which level there will be no significant effect on the viscosity of the mud. It has a further beneficial effect by promoting flocculation of drill cuttings.

Diamondback™

Reamer shoe from Weatherford. It recently proved its worth at the Abdaly Field in Kuwait: the case study is recorded on Weatherford's web pages. There it was applied in **directional drilling** of a bore of 7 inch diameter and reached a **TVD** of 3361 m having penetrated constrictions ('throats') and loose debris along the way. The inclination was 75 ° so, clearly, measured depth (MD) was:

3361 m/cos 75° = 13 km, and the **AHD** was

$$3361\text{ m} \times \tan 75° = 12.5\text{ km}$$

A general point which is a spin-off from this calculation is that the measured depth (MD), being the hypotenuse of the triangle, has to exceed the TVD. The latter is important to production, the former less so. The case study referred to reports: 'Well type: horizontal'. (*See* **RwC™**.)

Diamond Edge™

Series of drill bits from Varel International in Texas. Developed with **WOB** in mind, a **Diamond Edge™** bit is unsymmetrical when viewed along a line through its axis. Simulations and tests have shown that the absence of symmetry is a factor in vibration control during operation.

Dimlington

Gas terminal on the Yorkshire coast operated by BP, receiving gas by pipeline from several North Sea fields including **Ravenspurn North**, **West Sole** and Amethyst. It receives 'third-party gas' from a number of fields including **Babbage**.

DIPA

di-iso propanol amine. (*See* **ADIP process**.)

Directional drilling

Drilling in any direction other than vertical, including horizontal. **Directional drilling** can be achieved in one of two ways: by a pause in rotation of the drill string to reorientate the drill bit (sometimes referred to as 'sliding' it), or by a steerable arrangement in which drill bit reorientation can be accomplished whilst the drill string is still rotating. The 'direction' can of course be horizontal in which case the term 'horizontal drilling' applies. Directional drilling has the advantage of accessing more oil from a single drilling site and, conversely, enabling more oil to be raised at a single production site. In directional drilling a downhole motor can be used with a bent sub (deflection tool) in which case the trajectory is less likely to show discontinuities than with other means of such drilling. (*See also* **Downhole motor**; **Extended reach drilling**; **Sacha 171H well**; **TVD**; **Silver Bullet**; **Whipstock**; **Woodford OK**.)

Disconnectable turret

The forces on an **FPSO** with **turret mooring** have the same origin as those necessitating thruster operation in **dynamic positioning**. A 'turret FPSO' becomes unsafe if weather conditions are such that there is risk of breakage of the risers to it. Where such conditions prevail the FPSO is built so as to be disconnectable from the risers. This is achieved by making the risers integral with the mooring lines in what is known as riser turret mooring (RTM). In the event of disconnection the riser stays in position whilst the FPSO and its crew withdraw. An alternative approach is buoy turret mooring (BTM), in which a buoy forms a link in the mooring lines providing for disconnection.

With either type the structure from which the FPSO has been disconnected remains nearby and its use can be resumed when the hazards have passed. The turret has in fact found application in the North Sea, and at places including **Lufeng 13-1** where typhoons are frequent. It is also used where iceberg drift can occur, as at the **Terra Nova Field**. An **STP** facility is usually designed to be disconnectable. (*See* **Cossack Pioneer**.)

Discoverer Americas

Drill ship of the Transocean fleet having entered service in 2009. Built by Daewoo, it has **dynamic positioning** and can drill to depths in excess of 12 000 m in water depths up to 3500 m. The **POB** is 200 and its **moonpool** is 72.5 feet by 30 feet. It has recently been in use offshore Egypt and is expected to be relocated to the **Logan prospect**.

Discoverer Enterprise

Drill ship operated in the Gulf of Mexico by Transocean. Its **derrick** supports *two* sets of drilling works enabling two wells to be drilled from the vessel at once. It can work in water depths of 3000 m and has recently seen service in the Thunder Horse Field. (*See* **Jack/St. Malo Development**.)

Disko Island

Part of Greenland, in Baffin Bay which forms part of the west coast of the mainland and the scene of the first exploratory wells for oil in that country. They are currently being drilled by the UK firm Cairn Energy. The semi-submersible **Stena Don** and the drill ship **Stena Forth** are in position there. Wells having been drilled are the following. Alpha-1S1, drilled depth > 4000 m, has revealed oil over a **pay** of 'several hundreds of metres'[19] and, at a greater depth, oil over a pay of 400 m. *T8-1*, drilled to a **TVD** of 3250 m, revealed gas but was not deemed worth further investment and the well is now in plugged and abandoned status. Drilling at the well assigned the name *T4-1* continues. Whether the operations will go any further than exploration wells is not yet known. It was however reported in the *Guardian* newspaper on 22 March 2011 that Cairn Energy were seeking to sell some of their Greenland operations to one of the 'majors'. Write-off of the *T8-1* well already described cost Cairn Energy US$84.2 million. Shortly after the purchase of the reasonably promising *Alpha-1S1* well ConocoPhillips were awarded licences in Baffin Bay. Other companies with licences there include Shell and Dong Energy. (*See* **Leiv Eiriksson**.)

Doba Fields

In Chad, central Africa. The fields produce 120 000 barrels per day of crude oil. Chad is not a rich country, but did a few years ago become an oil exporting one, almost doubling its GDP immediately! The first export of oil from Chad's Doba fields was to the UK, where it was processed at the Pembroke refinery in South Wales. (*See* **Kome Kribi 1**.)

Dolphin Gas Project, Qatar

Non-associated gas field in the Khuff zone, at a water depth of 15 m to 40 m. **Spudding in** of the first well took place in 2001 and production began in 2007, there now being 24 wells at depths of about 3350 m in a carbonate formation. Peak production aimed for is 56 million cubic metres of gas per day. There is also condensate, about 100 000 barrels per day. There are two production platforms, and in late 2010 the operators of the project were inviting bids for a **jack-up** rig to house personnel in order that the two production platforms could undergo major maintenance. (*See also* **Bahrain Field**, **Yibal Field**.)

Double hull tankers, SRB in

Double hull tankers for oil are by now the norm, and use of single hull ones will become prohibited by law in EU and US waters in the near future. The reason of course is that in a collision the outer hull provides protection so that the tanks themselves are not impacted and broken open. Many species of **SRB** are thermophilic and, at temperatures up to 60 °C or higher, show a marked dependence of reproduction rate on temperature. The double hull configuration causes a 'thermos flask effect' by insulating the inner wall of the tanker, making its temperature higher than in a single hull vessel. This promotes SRB proliferation on the inner wall and hence corrosion. There have been examples of 'relatively new' double hull vessels showing significant corrosion attributed to this effect.

Douglas Field

Oil field in the eastern Irish Sea having commenced production in 1996. The water depth is a mere 33 m. The depth of the single production well is 650 m and the **pay** is 114 m. Associated gas is present but not abundant. The oil is high in sulphur compounds – hydrogen sulphide and mercaptans – making it 'sour'. There is desulphurisation offshore before transfer of the oil so treated to a terminal. (*See* **Liverpool Bay offshore storage installation**.)

Downhole motor

Placed in the drill string directly above the bit enabling bit deflection whilst the drill string remains in place. The downhole motor is used largely though not solely in **directional drilling**. For the purposes of this entry we consider the WellDrill™ range of downhole motors manufactured by Hunting Welltonic in Aberdeen. These come in a range of o.d. up to 2⅞ inches;

this size of WellDrill™ weighs 80 kg. A downhole motor does not drive the bit: it orientates it. The power requirements are correspondingly low. The WellBit™ under discussion has a power rating of 49 h.p. (36.5 kW) and can be used at **WOB** values up to 6500 lb (29 kN) and at bit rotation rates of up to 400 r.p.m. Like any other device at or near the **BHA** a downhole motor needs to be able to operate at temperatures of 200 °C or higher. The principle of *rotary* steering of a drill bit is that the bit rotation speed is adjusted so that the bit diverts from its previous trajectory. This is the alternative to use of a downhole motor in directional drilling. (*See also* **EZ-Pilot®**; **Spear drill bit**.)

DPFV

Dynamic Positioning Fall Pipe Vessel. (*See also* **Nordnes**; **Seahorse**; **Simon Stevin**.)

Drag bit

The *step type* drag bit has multiple cutting edges ('terraces' in some trade literature) on each wing of the bit structure, in contrast to the *Chevron type* drag bit which has three or four wings each providing a single cutting edge. **PDC**, diamond or **tungsten carbide** can be used on either type. Rotary speeds in the range 100 r.p.m.to 300 r.p.m. are typical of drag bits.

Drag reducing agent

A.k.a. a flow improver, introduced into crude oil or into refined material in small amounts (p.p.m. or tens thereof) to improve flow along a pipeline. Drag reducing agents are usually polymeric in structure and were first introduced by Conoco Phillips. Amongst such products currently available from that company is LiquidPower™. Its action is to reduce turbulence thereby redirecting internal motion – eddies and the like – to the direction of flow. It is mentioned in the entry for the **Trans-Alaska pipeline** that not all of the **pumping stations** will necessarily be in use at any one time, and this is because of the use of drag reducing agents. (*See also* **GO FLOW™ 126**; **RP™ II**.)

Dragon LNG Terminal

In Milford Haven, Wales, entering service in 2009. Companies involved include Petronas. The terminal is on a site previously occupied by a refinery. LNG delivered by tanker to the terminal is regasified and distributed by pipeline. Before departing Wales the gas goes to a **compressor station** near Swansea. The capacity of the terminal at commencement was about 2.2 million tonnes of LNG per year which becomes:

$$(2.2 \times 10^6 \times 10^3\text{kg}/0.016\ \text{kg mol}^{-1})/40\ \text{mol m}^{-3} =$$
3 billion cubic metres of gas per year

(*See* **Rudong LNG terminal**.)

Draugen Field

In the Norwegian sector of the North Sea. The water depth is 220 m to 295 m and the drilling depth 1650 m. It has a fixed platform[20]. There are 12 production wells and 2 **water injection** wells. More production wells are planned, to redress a recent decline in the production rate which currently stands at 63 000 barrels of oil per day with some associated gas. Oil from the platform is taken by pipeline to an **FSO**.

Drawworks

The device by means of which a drill in use in creating an oil well is raised or lowered. It has a very significant power requirement. As an example we consider *Discoverer Spirit*, a drilling rig operated by Transocean since 2000 having been built in Spain. It can drill in up to 3000 m of water. Its drawworks machinery is rated at 5000 h.p. (3.7 MW). Drawworks operation is of course intermittent. The drill itself will have a power requirement of the order of 1000 h.p. Power requirements for **dynamic positioning** are of the order of thousands of h.p. (*See* **Derrick**.)

Drill bit, life expectancy of ('bit life')

Taking an arbitrarily selected value from web sources for a drill with a bit made of **PDC**, the bit achieved 342 m of drilling at an **ROP** of 4.3 m per hour. Drill rotation rate is usually in the range 50 r.p.m. to 300 r.p.m. depending upon the nature and structure of the formation and the **WOB**. Simple related calculations follow.

Life span of the bit = 342/4.3 = 80 hours

Taking the mid range value for r.p.m.:

Total number of rotations = $175 \times 80 \times 60 = 0.8$ million

The above example was in the North Sea and the bit diameter was 8½ inches. Usually the longevity of a drill bit will be expressed as footage, and the record is believed to be 22 930 feet (6989 m, exceeding that in the above example by a factor of exactly 20) at the **Zuata Field**. That was with an 8½ inch PDC drill bit manufactured by Hughes Christensen. The 22 930 feet was not all vertical drilling; there were also segments connected to the main well in a 'fishbone' structure. The record for a tricone bit is 220 514 feet (6257 m) at the **Al Shaheen Field** in Qatar. The life expectancy of a **drill collar** is more predictable than that of a drill bit because it has no direct experience of the erosion due to the formation. One would expect that a drill collar would see multiple 'bit lives' and a single 'bit life' might well be over several operations and therefore be cumulative. An ROP value of 665 feet per hour (203 metres per hour) offshore China in 1999 is said to have a claim to a place in the record books. That is two orders of magnitude higher than the value in the above calculation and was also with a PDC bit. (*See also* **Dull bit**; **Tiber field**.)

Drill bit, largest in the world

In the entry for **spudding in** a value of 'two feet or more' is given for the diameter of a drill bit for use in that operation. In fact the largest drill bit in the world, now on display in a museum in Norway, is only 3 feet in diameter. It weighs 1.8 tonnes and has a **bicone** structure. The largest drill bit currently manufactured by National Oilwell Varco, one of the oldest and biggest manufacturers of drill bits in the world, is 17½ inches in diameter.

Drill collar

Part of the **BHA**. For the purposes of description those manufactured by the oil field service company National Oilwell Varco (NOV), whose HQ are in Houston, have been selected. Having the brand name Grant Prideco (an independent company before acquisition by NOV), these are made of steel of design stress about 750 MPa, and are available with or without spiralling of the outside surface. Lengths are 30 feet or 31 feet, and diameters in the range 3⅛ inch to 11 inch are available. Steel drill collars from International Drilling Services in the UK come in lengths and in the same diameter range. Drill collars made of Monel® are also widely used and these, unlike steel drill collars, will not cause magnetic interference with **MWD** instruments. Alignment of the axis of a drill collar with that of what precedes it in the downhole structure will never be perfect and the consequent imperfection is referred to as eccentricity. It leads to vibration and possible eventual failure of the drilling operation. The life expectancy of a steel drill collar can be raised by hardbanding, by which is meant use of a layer of **tungsten carbide** along part of the outside surface.

Drill ships, newly constructed

The table below gives details of a number of drill ships expected to enter initial service in 2010/2011.

Name	Details
Norbe X	Built by Daewoo. Sea depths up to 3000 m.

Name	Details
Cerrado	Built by Samsung. Sea depths up to 3000 m. **Drawworks** rated at 5750 h.p. Top drive drilling.
Deep Ocean Molokai	Built by Samsung. Sea depths up to 3600 m. **Derrick** hoisting capacity > 900 tonnes.
Carolina	Built by Daewoo. Sea depths up to 3000 m. Drawworks rated at 4500 h.p. Derrick hoisting capacity > 900 tonnes.
Ocean Rig Corcovado	Built by Samsung. Sea depths up to 3000 m. Drawworks rated at 4290 h.p.

All of the drill ships in the table are 'competitive rigs', that is, they are available for use anywhere by an operator who makes an acceptable bid. They all have **dynamic positioning**. A drill ship at the scene of drilling has 'drilling' status: one being taken from one scene of drilling to another has 'en route' status. (*See also* **Ben Avon**; **Discoverer Americas**; **Jasper Explorer**; **Noble Roger Eason**; **Stena Forth**; **West Navigator**.)

Drill string shock absorber

Vibrations during well bore drilling lead to accelerated wear of drill bits and other components and impairment of the **ROP**. A shock absorber can be used in dampening such vibrations. The shock absorber, which will of course contain a hydraulic fluid, might be placed at any position along the drill string but in general the closer to the drill bit it is the greater the effectiveness. The most advanced such device is the **AVD™**.

Drilling at zero WOB

Already mentioned in the entry for **Bit booster®**, the only further information on this which the author has been able to access is from a blog in which impressions on field trials are reported. The trials were of very limited scope. One reported impression is that the **ROP** is at worst not adversely affected, possibly slightly raised. Another contributor to the blog recounts his experience with the approach when drilling into a very soft formation in the Gulf of Thailand.

Drilling fluid

Synonym for drilling mud. Such 'muds' can be water based or oil based. The former can have either freshwater or seawater as the base. The latter will contain a base oil such as **ODC-15™** or an **LTMO** with further constituents (e.g., 'viscosifiers') incorporated to give the required characteristics. The term emulsion is used when there are both organic and aqueous phases as with an **invert emulsion**. Increasingly important are synthetic base muds (SBM[21]), for example **RHEO-LOGIC™**. In these the organic phase is not a petroleum distillate or residue but compounds of precise composition synthesised from them. Drilling fluids can encounter temperatures of 300 °C or even higher. Filtration control is more likely to be an issue at high temperatures and pressures. Water based drilling fluids are usually on the alkaline side of neutral, a pH range of 8 to 10.5 being expected. Obviously some applications will require a customised drilling fluid. For example, underbalanced drilling means drilling in such a way that the pressure applied to the formation being drilled is lower than that inside the formation. In conventional drilling the opposite is so. The 'underbalance' is achieved by use of a low-density drilling fluid and the method can be applied either to rotary drilling or to **coiled tubing drilling**. When circumstances are underbalanced oil is drawn from the formation towards the drill (*See also* **Swellable elastomer**; **UBCTD**.)

Drilling jar

Device for removing something stuck in a bore being drilled therefore preventing advance of the drill bit. Being integral to the drill string it draws on the same source of energy as the drill string itself. When its 'jar' function is needed, it applies a force ('detent force') which, it is intended, will remove the stuck object. Positioning of a drilling jar along the drill string is obviously a factor in effectiveness and 'unit length' will be the length of a **drill collar**. A drill string close to the **BHA** has a *BHA stabiliser* which prevents deviations of the drilling trajectory close to the BHA. Manufacturers of drilling jars will typically specify positioning 'two drill collars above the BHA stabiliser'. Temperatures of operation close to the BHA will often be as high as 150 °C. (*See also* **Sledgehammer™**; **Sup-R-Jar®**.)

Droshky Field

Oil and gas field in the Gulf of Mexico, productive since 2010. The water depth is 914 m. Drilling of the first two production wells was by the **semi-submersible rig** Henry Goodrich, a member of the Transocean fleet. Each well was a sidetrack from an existing well casing arrangement. **Pay** encountered was 180 m for one of the wells and 90 m for the other. Two more wells were drilled by the semi-submersible rig Paul Romano, operated by Noble. For production purposes the wells were tied back to Bullwinkle, a **tension leg platform** owned by Shell.

DSS-20-CAS-M

Semi-submersible rig built in Singapore, able to operate in 1000 m of water and to drill 9 km below the sea-floor. Its **derrick** is at a height of 55 m and can take a load of 900 tonnes. It has two pontoons in a catamaran structure.

Dual-diameter pigging

A dual-diameter **pig** consists of two axially linked mechanical pigs of different diameters, for example 6 inches and 8 inches. The larger will always be a **foam pig**. When operating at 8 inch diameter the 6 inch pig will be redundant. When operating at 6 inch diameter the 8 inch pig will compress but will, on exit, return to its original size and perform as previously on an 8 inch pipe. Such a pig is a 6-8 pig, and it is expected that the larger pig will after compression return almost precisely to its original size. This is on the basis of compression relief in going from 6 inches to 8 inches and one would not expect the same response or effect on going from 6 inches to 7 inches. The pig is therefore for two diameters only and not intermediate ones. A pig which *can* operate across a range and is not restricted to two diameters is a *multi-diameter* pig. There are modern dual-diameter pigs which have diameter separations much larger than the two inches in this example. An alternative to foam for the pig which has to adjust is a cup made of a material such that it can restore its configuration after being squeezed; polyester urethane is a common choice. (*See* **Åsgard pipeline**.)

Dukhan Field

Onshore field in Qatar, producing since 1940. It currently produces up to 335 000 barrels of oil per day, 'up to' because at a field which has been in production for so long well condition will restrict production. As at the **Bahrain Field** there are wells producing non-associated gas, 58 of them compared with 300 oil wells. There are also wells for **water injection**. The oil is taken by pipeline to the port city Umm Said, where there are refining and exporting facilities.

Dull bit

Drill bit having become unserviceable, either not working at all or giving a lower than required **ROP**. Having regard to the facts that (i) removal of the drill string to replace a bit is a costly operation and (ii) the bit itself is a significant capital asset, there will be an examination of a dull bit and a report on it in which **IADC** terminology and symbols will be used. Tooth condition is reported on a linear scale from 0 to 8, 0 signifying no loss of tooth height and 8 complete loss. Pairs of letters apply to features that can be judged from visual observation. They include BC for broken cone, OC for off-centre wear, ER for erosion, CC for cracked cone and LC for lost cone. (*See* **Drill bit, life expectancy of ('bit life')**.)

Duramex™

Elastomer material developed by Baker Hughes for use in downhole vibration damping. It has been used with **downhole motors** made by the same company to the enhancement of performance, for example in drilling for 'tight gas' at Pinedale, WY.

Durango well

Well in the North Sea off the English coast, productive of gas which is taken to the Waveney platform nine miles away in what might reasonably be called a **long-distance subsea tieback** operation (though the distance to the host platform is less than a third that at the **Cottonwood Field**). Gas is taken from there to the terminal at Bacton off the Norfolk coast. At Durango the **pay** is at a depth of 2750 m below the sea.

DURATONE® E

Drilling mud **filtration** control agent from Halliburton made from suitably pre-treated **leonardite**, a 'lignitic product' in the terminology of **drilling fluids**. Directed at **invert emulsion** systems, it comprises about half of one per cent of the weight of a mud into which it is incorporated and has a secondary beneficial effect in helping to stabilise the emulsion.

Dynamic positioning

An alternative to anchorage of oil rigs, and common in deeper water applications. 'Thrusters' provide as necessary equal and opposite influences to those which would have caused the rig to drift in water. The power ratings of the thrusters are major; for example on Discoverer Seven Seas, built by Mitsui and operated by Transocean, there are six thrusters each of 2500 h.p. A production vessel with only single point mooring might require dynamic positioning for 'station keeping'. (*See also* **Chikyu**; **Glomar Explorer**; **GSF Jack Ryan**; **Gusto CTV**; **Jasper Explorer**; **LB 200**; **Production rigs, propulsion of**; **Rubicon Offshore, vessels operated by**; **Munin**; **Noble Leo Segerius**; **Noble Phoenix**; **OceanCourage**; **Peter Schelte**; **Rollingstone**; **Saipem 10000**; **Sarah**; **Seahorse**; **Shuttle tanker**; **Skandi Aker**; **Solitaire**; **Stena Don**; **Stena Forth**; **Well Enhancer**.)

Dzheitune field

In the Caspian Sea offshore Turkmenistan. It is being developed by Dragon Oil (HQ in the UK). The **jack-up** rigs **Astra** and **Iran Khazar** have both been in use there. The water depth at the site of drilling is as low as 8 m. In January 2011 it was announced by Dragon that a development well assigned the name B/150 had in testing produced oil and that a second well, B/153, had undergone **spudding in**. The space between B/150 and B/153 was such that movement of the jack-up rig from one to the other was by skidding. (*See also* **Bagaja gas condensate field**; **Dzhygalybeg field**.)

Dzhygalybeg field

Offshore Turkmenistan productive of oil and gas, water depth 42 m. The operator is Dragon Oil and drilling of new wells will be from a platform called Dzhygalybeg A. It is expected that this will commence operations in early 2012 and the initial drilling programme is eight wells. Although development is the keynote at Dzhygalybeg and at the **Dzheitune Field** there is already limited production, currently 57 000 barrels per day between the two. (*See* **Bagaja gas condensate field**.)

East Baghdad Field

Productive of oil since 1980. Parts of this field are in the Zubair sandstone formation, in which permeabilities average 700 millidarcy and porosities are up to 30%. Current production at the field is about 15 000 barrels per day, but its reserves are such that it could be very much higher than that and foreign investment is what stands between the current production level and a level two to three times higher.

Easywell

Swell packer from Halliburton. It was recently used at the **Andrew Field** in the North Sea.

Eclipse Wireline

Manufacturers of trucks for **slickline** operations, HQ in Logan UT with a presence in the UAE. The company use aluminium cab bodies which, within a range, the customer can select. The truck once constructed and commissioned can hold 25 000 feet of slickline in the diameter range 0.108 inches to 0.125 inches and an equivalent length of **braided line** of 7/32 (=0.22) inches. It can deliver slickline from a drum at rates between 10 feet and 2250 feet per minute.

Eidesvik Offshore

Operator of vessels for the offshore oil and gas industry. The company has recently introduced a support vessel powered by LNG. Four such vessels are in service and a fifth on order. This will be built in Norway and will have dual engines made by Wärtsilä.

Ekofisk Field

In the Norwegian sector of the North Sea, discovered in 1969 and producing without interruption since 1971. The water depth is about 70 m. The reservoir formation is chalk, of low permeability. Over one period of the field's production life there was **depletion drive**, later abandoned in favour of **water injection**. Well depths are in the range 2900 m to 3250 m. By now there are many plugged and abandoned wells and considerable decommissioned infrastructure. By the same token new infrastructure is being put in place. Currently there are about 25 platforms. The fact that the formation is chalk leads to downrating of platform support because of higher susceptibility to subsidence. The pile assembly which holds in place Ekofisk2/4X, one of the major platforms at the field, is in 76 m of water having a formal rating for up to 90 m. Ekofisk 2/4X can receive from production wells 0.26 million barrels of oil per day. This is a little under half the total production of the field. Some of the 'platforms' are not for production, but for operations such as water injection. Ekofisk 2/4W is for water injection, the oil it so raises going to Ekofisk 2/4A which is a wellhead platform. Ekofisk 2/4W was hit by the oil field services vessel Big Orange XVIII in 2009: this obviously seriously affected production at 2/4A. (See **Nitrate ions, action on by SRB**)

Electric submersible pump

Widely used both at onshore and offshore fields, this device uses a pump of centrifugal design with variable impeller speed powered by an AC motor. The electricity requirements of electric submersible pumps are typically 3 kV to 5 kV. Examples of the use of the electric submersible pump (ESP) at oil fields are very many and a few examples are given in the table below.

Location and operator	Details
Otter (formerly Wendy) Field, North Sea. TotalElfFina	Transfer of oil from a distant (23 km) platform – the Eider platform – by downhole ESPs. The previous record for such movement in the North Sea 14 km. Two ESPs in each of three wells at the Otter Field. Such a transfer called 'subsea step-out'.
Parque das Conchas, offshore Brazil. Shell	Production at a water depth of 1780 m. Oil transfer to an **FPSO** by six ESPs each of 1500 h.p. rating. Also a cyclonic subsea oil-gas separator.
Rumaila oil field, Iraq. BP	Plans to install 50 ESPs.
Talco Field, TX. Exxon	Field productive for over 70 years. An extra 114 000 barrels per well realised by the installation of ESPs.

Location and operator	Details
Stage Field, NW Shelf, Australia. Apache Corporation	An ESP in each of the five production wells.
Manifa Field, Arabian Gulf. Saudi Aramco	ESPs at production wells in shallow water.
Ofa Field, Nigeria. Afren	Use of as ESP to supplement reservoir pressure in well evaluation.

The ESPs in service at the Otter field are advancing production in two ways. They eliminate the need for an offloading facility (such as an **FSO**) at Otter and they eke out more use from the Eider which was in fact scheduled for shutdown until this link to the Otter field was conceived. The combined 90 000 h.p. rating of the ESPs at Parque das Conchas converts to 6.7 MW requiring about 20 MW of electricity. This is generated at the **FPSO**, mooring of which is by catenary means. The proposed ESPs at Iraq's massive Rumaila field, which is being brought back into service concurrently with the **Bazergan Field** under a contract awarded to Weatherford, are part of an overhaul of a group of wells which are declining in production rate. Amounts accessible are increased by ESP usage, and this is also true of the **Talco field** (following row). At the Stag Field there is a **fixed platform**. The Manifa Field (row six) is under redevelopment and is expected to be producing half a million barrels a day by 2013, also associated gas and condensate. (*See also* **Buckland Field**; **East Baghdad Field**; **Helena**; **PI**; **Rajasthan block**.)

Elm Coulee Field

In eastern Montana, discovered in 2000 and by 2007 one of the 20 most productive oil fields in the US. This rapid growth is largely due to the proliferation of horizontal wells at the field from its beginning: there are now over 350 such wells at Elm Coulee. The reservoir is of only moderate **porosity** and permeability, 8% to 12% and 0.05 millidarcy to 0.1 millidarcys respectively. Current production is about 60 000 barrels of crude oil per day.

Elswick Field

Small onshore natural gas field in western Lancashire, operated by Warwick Energy and producing since 1996. The gas is not pipelined but burnt on site, making electricity at about 1 MW for the grid using not a gas or steam turbine but, appropriately to the scale, an engine. Also operated by Warwick

Energy was a 9 MW power generation facility in the English midlands which used natural gas from the now depleted Ironville Field.

ELWC

Economy lightweight ceramic, term applied to some **proppants**.

Emeraude Field

Offshore the Republic of Congo, water depth 250 m to 400 m. Having become moribund, it was revived in 2009 by the UK based company Perenco who installed a drilling and production platform in the form of an **MOAB** at the central part of the field. This was preceded by major well stimulation endeavours. Oil production is currently of the order of 2500 barrels per day. The oil from the field is more than usually viscous, and this is being addressed as necessary by steam injection. Perenco operate three other fields in Republic of Congo waters. One is Yombo, where an **FPSO** is in position and 10 000 barrels per day are being produced. The others are Likouala, producing 6000 barrels per day, and Marine IV which is currently being appraised. (*See* **Conkouati**.)

Emlichheim

Oil field in Germany operated by Wintershall. Over a 55-year period annual oil production at Emlichheim was a steady one million barrels. For some time steam injection has been used to increase production at the field. In general steam injection has two effects: raising of reservoir pressure and reduction of oil viscosity. In any particular application one or other might be more important, and a reader of the Wintershall web pages concludes that it is the viscosity effect which is seen as being important at Emlichheim. Increased production at Emlichheim is in progress by way of horizontal drilling from an existing vertical well. This well is 800 m deep, but the horizontal sidetrack will occur at a depth of 500 m into it, and will extend about 800 m from there to access parts of the reservoir not previously drawn on. (*See* **Aitingen**.)

Emulsion breaker

The separation of oil from water is natural because of the immiscibility of the two except where they have formed an emulsion. A primary emulsion breaker is then used to separate oil from water. Water having entered a separator from a well itself needs cleansing and for this a reverse emulsion breaker is used. Sand and mineral particles also need separating and this requires a wetting agent. Where the crude is acidic the emulsion breaker might also have a role in pH adjustment of the residual water.

ENOC

Emirates National Oil Company. (*See* **Jebel Ali Refinery**.)

Ensco 70

A further example (see **GSF Britannia**, **Maersk Guardian** and **Noble Al White**) of a long-in-the-tooth rig. Built in 1981 by Hitachi, it underwent a major upgrade in 1996 and is based in the UK sector of the North Sea where it operates in water depths up to 250 m. It is one of a family of **jack-up** facilities bearing the Ensco name. Another is Ensco 101 which, with a US flag, also currently operates in UK waters. This was built in 2000 in Singapore for work in water depths up to 400 m. (*See also* **Al Ghallan**; **Arthit Field**; **Big Foot Field**; **Hanze F2A Field**; **South Tor Pod**; **Spud can**.)

Ensco 100

This **jack-up** rig recently carried out a hydraulic fracture operation at the **Victoria gas field** the southern North Sea, whereas a **semi-submersible rig** is more conventional in hydraulic fracture. An exploration well at the field had given promising results about a year earlier. The hydraulic fracture used 259 577 lb (120 tonnes approx.) of **proppant**. Formulation of the **hydraulic fracturing fluid** was done in the 'rig pits' on the jack-up, the intended use of which is storage of drilling mud. The field is now producing, the gas being taken to a terminal on the Lincolnshire coast. A record is claimed that the hydraulic fracture was the largest 'rig based frac' ever to have been conducted in the North Sea. A 'frac' from a semi-submersible rig would by analogous semantics be 'vessel based'[22]. (*See also* **Arthit Field**; **Jasmine Feld**.)

Equatorial Guinea LNG Plant

On Bioko Island, having at present a single LNG train ('Train 1') which has a claim to uniqueness in the following way. The topography of the site is such that pipeline passage of LNG to tankers involves a drop in altitude of 60 m, and early suggestions that this might be accommodated by clearance of vegetation and installation of supports met with rejection as the ground is too unstable. Instead a suspension bridge was built to support the pipelines conveying the cryogen. This is believed to be the only such arrangement anywhere. Plans for at least one more LNG train at the plant are under way. (*See* **North Adriatic LNG Terminal**.)

Erha Field

Offshore Nigeria, water depth about 1200 m. It is operated by Shell and ExxonMobil. There are 32 wells in all – production, **water injection** and gas reinjection – and an **FPSO** having the same name as the field and carrying the Nigerian flag is in long-term service. Production of oil is 150 000 barrels per day and of associated gas 8 million cubic metres per day. The FPSO is one of the largest in the world. A new build, it has **spread mooring** and can hold 2.2 million barrels of oil. It can receive 210 000 barrels per day, enabling it to accommodate a substantial increase in the production. Enough associated gas for the vessel's own requirements is dehydrated on board, and the rest reinjected into the reservoir.

Etame Field

Oil field, with significant associated gas, offshore Gabon. The sandstone formation varies in its characteristics. However, permeabilities are as high as 3000 millidarcy in places, and **porosity** values are up to 30%. There are currently five production wells, oil from which goes to the **FPSO** Petroleo Nautipa. This is a fairly elderly facility, having been built in Japan in 1975. It has been at the Etame Field since production began there in 2002 and has **spread mooring**. Etame is a small field, estimated reserves 23 million barrels. Petroleo Nautipa has a current annual throughput of six million barrels making the life expectancy of the field less than four years unless there are new discoveries that can be tied back to its infrastructure. (*See also* **Acergy Condor**; **Cusiana Field**; **Fulmar Field**; **Olowi Field**.)

ETAP

Eastern Trough Area Project. (*See also* **Columbus Field**; **Lomond Field**; **Machar Field**; **Marnock**; **Skua Field, North Sea**.)

European countries with nil or negligible oil production, examples of

These are given in tabular from below with comments. Import figures for oil are for 2010.

Finland, population 5.4 millions. EU since 1995.	337 900 barrels per day imported from countries including Norway, Russia, UK. Extensive downstream structure with Neste Petroleum.
Latvia, population 2.2 millions. EU since 2004.	43 400 barrels per day imported, more than half of it from Russia.
Luxembourg, population 0.5 millions. EU since 1958.	59 210 barrels per day in the form of refined products. No *crude* oil import because no refining capacity. Kerosene supply for the airport via a pipeline.

Switzerland, population 7.9 millions. Non EU.	269 400 barrels per day including both crude oil and imported refined products. Some refining capacity for the crude oil.

Two of the four countries that feature in the above table are amongst the sixteen EU countries which, like the US, have a strategic petroleum reserve: Finland and Luxembourg.

Everest Field

Condensate field in the North Sea. A **pipeline repair clamp** of split sleeve design was recently supplied there, NOT for immediate use but for possible future use. The pipeline to which it will be applied if needed is of o.d. 34.6 cm and the diameter of the clamp once the sleeve has been closed is 36 cm. The difference of course is due to allowance for the thickness of the fluoro-elastomer present. The clamp was made by Furmanite (HQ in Richardson TX) and the idea is that if there is a breakage in the pipeline taking condensate from Everest it can be fitted at short notice simply by bolting: there being no welding needed.

Extended reach drilling

There is no non-arbitrary definition of this in terms of depth, 'reach' or whatever. It means drilling having the effect of making oil over a wide area accessible for transfer to a single production facility. Clearly it enables a good return on investment to be obtained from infrasructure. Extended reach drilling to a subsea reservoir can be shore based as at the **Chayvo Field**, where ten wells are being drilled from a land rig on the beach. Distances from beach to well are up to 11 km. (*See also* **AHD**; **Armada platform**; **Nikaitchuq Field**.)

External turret

Such a device is not integral to the hull structure of an **FPSO** or an **FSO** but attached to the front of it (the 'fore'), otherwise operation is as for **turret mooring**. It is a preferred method in shallower waters, for example at the **Yoho Field**. (*See also* **Farwah**; **Fluminense**; **Kikeh**; **Pathumabaha**.)

ExxonMobil Fast-Drill Process

Having been initially tested at a Texan oil field, this process has since proved its worth at other places, including Sakhalin. As drilling proceeds the energy being expended at the drill bit face is measured and recorded on a digital display. The driller is therefore alerted when this energy increases. The hardness of a rock structure seldom changes in a discontinuous fashion so an abrupt change in the energy used at the bit face is likely to be indicative of some other factor, most probably clogging of the drill bit teeth, in which case in situ cleaning of the bit with **drilling fluid** is possible. Energy that would have been lost by drilling with a blunt bit is thus redirected to the enhancement of the **ROP**. The ultimate consequence of continued operation with a clogged drill bit is a decline in the ROP to unacceptable levels, in which case withdrawal of the bit for replacement is necessary. This adds a great deal to the cost of the drilling process.

EZ-Pilot®

Downhole tool for rotary steering of a drill bit, a Halliburton product. Also from Halliburton is the FX Geo-Pilot® range of drill bits for use in directional drilling either with a **downhole motor** or by rotary means.

EZReam™

Reaming tool from Baker Hughes. Its steel supporting structure will accommodate a **PDC** or a **roller cone drill bit** as the cutting surface. Its value is in getting to the **TVD** after a borehole has become severely narrowed. This is always the ultimate goal in **reaming**, and failure to attain it will entail re-decision with regard to the depth of the **pay**. (*See also* **Borrox®**; **GaugePro XPR™**.)

F21 Well (BP)

At Wytch Farm, on the south coast of England. Wytch Farm is the largest onshore oil field in Europe. When it was discovered in the early 1970s its value was doubtful because it is on top of some of the most scenic parts of England containing real estate of superlatively high value. Vertical drilling was totally out of the question and **directional drilling** techniques were largely developed at and for Wytch farm. At Wytch farm in 1994 directional drilling across a measured distance of 11 225 feet was accomplished in a single drilling run lasting 27 hours, giving an **ROP** of

$$11225/27 = 416 \text{ feet per hour}$$

A **PDC** drill bit was used and there was a **downhole motor** for bit orientation. This is about an order of magnitude higher than ROP values given elsewhere in this volume and acceptance of it requires that other values of comparable magnitude be noted. An ROP of 665 feet per hour has been documented at the Bohai Bay offshore China with a PDC drill bit, and a value of 554 feet per hour at the Al Shaheen Field in Qatar with a Hughes **MX** drill bit. ROP values as high as this are exceedingly rare and are beyond the capacity even of the **ExxonMobil Fast-Drill Process**. There are great difficulties in asserting that any one ROP value is a 'record' (although some blogs do). The ROP depends *inter alia* on the drill bit and on the directionality of drilling, and on the many factors inter-related with those. To state that a particular ROP is a 'record' is unsound because of lack of a benchmark unifying these many influences.

Fall pipe vessel

Vessel for **subsea rock installation**. (*See also* **Nordnes**; **Rocknes**; **Rollingstone**; **Sea Horse**; **Side dumping**.)

Farha South-5

Well in Oman developed by the Swedish Tethys Oil. It contains a horizontal section of length 1600 m and an **electric submersible pump**. The oil is light, being of 44 degrees on the API scale (density 806 kg m^{-3}).

Farwah

FPSO operated by Total, a new build. It entered service in 2001. Moored to an **external turret** with six mooring lines attached to anchors, it is in long-term service at the **Al Jurf Field**. It can receive up to 35 000 barrels of oil per day. A vessel from the same 'stable' is Girrasol, also an FPSO and active on behalf of Total in Angolan waters. (*See also* **Rosa Field**.)

Fateh Field

Offshore Dubai, discovered in 1966 and the first of Dubai's oil fields to enter production. It is operated by the Dubai Petroleum Company (DPC). It is most noted for being the scene of the first use of a wholly new design of oil storage tank called Khazzan 1, which was lowered into position in 47 m of water in 1969. Khazzan 1 had a bottomless structure: it was always full, either with oil, seawater or both. It worked on the principle that oil admitted at the top displaces water from the base and that crude oil, being less dense than water and immiscible with it, would always occupy the upper portion of the enclosure. The capacity of Khazzan 1 was half a million barrels. In 1970 export of oil from Fateh began, initially to Japan.

Fault

When applied to oil and gas production, this term means a part of the geological formation where there has been displacement of rock by prior seismic activity. In well drilling it can make for difficulties. If the fault is not along the drilling trajectory but within the reservoir, recovery can be affected as at the **Njord Field**. Drilling can exacerbate a fault distant from the drilling site. Very importantly, 'fault' should not necessarily be taken to have a negative meaning in oil and gas production. In exploration, discovery of a 'fault' can be auspicious in terms of there being oil. One often reads that an oil field is in, or at the edge of, a particular 'fault' in which case the term 'fault block' might be used. This is in fact true of the **Visund Field** amongst many others. One also encounters the term 'growth fault', most commonly in relation to N. American fields. (*See also* **Meji Field**; **RwC™**; **Singleton Field**; **Throw**.)

FBE (*See* **Fusion Bonded Epoxy**.)

Fusion bonded epoxy, a substance with good anti-corrosion characteristics used in the insulation of subsea pipelines.

Filtration

In oil well terminology, penetration of a rock formation by drilling mud. An equivalent term is 'fluid loss'. This is an unwanted effect for a number of reasons. It denudes the drilling mud of some of its constituents and can lead to bore blockage. Filtrate in the pores makes the rock hydrophilic and this can cause it to attract some of the water upon which a drill mud function depends. On the other hand a thin layer of such filtrate can sometimes be beneficial in preventing 'fluid loss', and such a layer is called filter cake. Depending on the composition and nature of the mud, filter cake can develop to a thickness such that it restricts drilling in which case filter cake removal, for example by use of **Arcasolve™**, is required. Avoidance of filtration can be achieved by a fluid loss additive such as **FLR-150/170** or **mud acid**.

Firenze

FPSO having over a long period been in service in the Aquila Field in the Adriatic Sea where production began in 1998. The owner and operator of the field is Eni. Initially production was from two horizontal wells and the production rate was up to 17 000 barrels per day. There has been expansion at the field with the result that Firenze has been replaced by a bigger FPSO, which was a rebuild.

FIV

Formation isolation valve. (*See* **Arcasolve™**.)

Fixed platform

In contrast to a **spar platform** or a **tension leg platform** in either of which the platform is moored to the sea-floor by cables or chains, a fixed platform is attached to the sea-floor by legs constructed of steel and concrete. It is sometimes possible for the total weight to be sufficient to keep the platform stationary without any additional device, in which case it is important to have a good area of contact between the steel/concrete support and the sea-floor. If such self-support is not possible, piles into the sea-floor provide foundations. The fixed platform finds application, not to the exclusion of other approaches, where water depths are not in excess of about 500 m. Piper Alpha, the destruction of which by fire in 1988 caused the loss of 167 lives, was a fixed platform. Its sea depth was only 140 m. It is the rigidity of the legs which precludes fixed platforms in deeper waters. Isolation of the legs from each other – a factor in the depth limitation – can be avoided by having a structure in which the vertical supports have a tower structure with components additional to the vertical ones providing for some flexibility. This arrangement is called a compliant tower[23]. A major facility in current use having the compliant tower design is **Petronius A**. The first compliant tower was Baldpate in the Gulf of Mexico in exactly 500 m of water. At Baldpate the distance from the sea-floor to the tip of the flare on the platform is 580 m, making it one of the highest free-standing structures in the world (cf. Empire State Building, 443 m to the spike of the lightning conductor). (*See also* **Angel Field**; **Blane Field**; **Brage Field**; **Gannet Field**; **GBS**; **Lomond Field**; **Oseberg Field**; **Thebaud platform**; **Tombua-Landana**.)

FLEX LNG

Based in the British Virgin Islands with offices in London and Oslo, this company's business is the development of vessels including **FNGU**s. These are mobile and so will be able to move from one scene of natural gas production and/or storage (for example an **FPSO**) to another. The LNG plant at the **Snøhvit LNG facility** is not mobile, nor is it really 'offshore' as it is integral to the operations at **Melkøya Island**. The enterprise will therefore have some claim to being the world's first 'offshore LNG facility'. Note that FLEX LNG are involved in design, conception and development: the building of their vessels is contracted out and the company have in fact placed orders for four FNGUs with Samsung. Obviously an FNGU can be put to use at a 'stranded field'. This is one factor in the assertion by FLEX LNG that use of the FNGU will make for lower capital expenditure per unit LNG than transfer of offshore gas to onshore liquefaction facilities. (*See* **PTT PCL**.)

Floatec

Planned **tension leg platform** for the Papa Terra Field in the Campos Basin, the first such platform in Brazilian waters. The company Oil States, at its facility in Aberdeen, Scotland, will manufacture the tendon connectors, both those for attachment to the sea-floor and those for attachment to the platform. The international HQ of Oil States is in Arlington TX. Oil risers at Floatec are of the **top tension(ed)** type. (See **PWD**.)

Flora Field

Oil field in the North Sea, producing since 1998. Peak production over that time has been 9700 barrels per day. There are two horizontal production wells and one well for **water injection**. There is tie-back to **Uisge Gorm** at the Fife Field.

FLR-150/170

Fluid loss additives from Weatherford. Each contains particles that will block the channels through which fluid loss can take place without jeopardising the permeability because the particles, being hydrocarbon based, dissolve in the oil being released. The only difference between the two is that the 170 form can operate at temperatures about 20 °C higher than the ceiling temperature for the 150 product. There are a number of other such products from Weatherford, including FL-300HT. It was expressly developed for formations with low permeabilities – down to 1 *micro*darcy – and contains starch particles which will naturally degrade and might otherwise have made the permeability even lower.

Fluminense

FPSO in service offshore Brazil moored by an **external turret**. Lines to the turret are made of polyester and are tensioned, in contrast to catenary mooring which is conventional with the external turret. Consequently, Fluminense has been able to operate in water depths of 800 m, a record for external turret mooring.

FNGU

Floating Natural Gas Unit. (*See* **FLEX LNG**)

Foam pig

Made from polymer foam often polyurethane, having some of the rigidity of a mechanical pig and some of the penetration capacity of the **gel pig**. They come in bulk densities in the approximate range 15 $kg\,m^{-3}$ to 150 $kg\,m^{-3}$ and will be capable of travelling two to three hundred miles along a pipeline before ceasing to function. An example is the InPipe range, manufactured in the UK. This range 'starts' at simple soft foam of 32 $kg\,m^{-3}$ density whose applications include 'proving'. By this is meant passage along a pipeline which has never before been 'pigged' because pigging was not factored into its design. Successful passage of a pig on a 'proving' mission along such a pipe makes for confidence that a more advanced pig of some sort will be applicable at the pipeline. Use of a basic foam pig followed by the use of more complex ones is sometimes referred to as progressive pigging. A simple foam pig like that under discussion also has swabbing (removal of contaminants from the inside pipe surface) and drying roles. Developments of the basic foam pig, for example coating with a silicon carbide abrasive, are common. Other such accessories include a spirally positioned (by means of straps) brush on the pig, and the brush may be light- medium- or heavy-duty. Alternatively, the foam material itself might have a spiral. (*See also* **Dual-diameter pigging**; **Omnithane®**.)

Foinaven Field

Oil field (estimated reserves 380 million barrels) with associated gas (estimated 6 billion cubic metres) off the Shetland Islands. The operator is BP. An **FPSO** (Petrojarl IV) is permanently positioned at the field, which has been producing since 2000. Oil from the FPSO is offloaded on to **shuttle**

tankers. The water depth is about 450 m and field development required **extended reach drilling**.

Formates

Use of brines containing potassium formate (KCOOH), sodium formate (NaCOOH) or caesium formate (CsCOOH) in water based drilling muds and in **completion** fluids is increasing. For example, at the Khuff Field in Saudi Arabia drilling muds with a formate solution base have been used, and in India's Thar desert, completion fluids with a formate solution base. Aqueous solutions of formates are of higher pH than water, making them suitable for use as completion fluids or as drill-in fluids where the formation is sensitive to acid.

Formation structure, hardness of on the Mohs scale

Like other physical properties including **porosity**, the hardness of a formation will depend on conditions. The dominant mineral in sandstone, quartz, is typically 6–7 on the scale while carbonate formations are predominantly composed of calcite which is 3 on Mohs scale. The salt formation at **Lula Field** contains minerals of a Mohs value similar to calcite. Even so, at a salt formation **ROP** values in drilling averaged over the drilling period tend to be low through frequent periods of **NPT**. This is because salt occurs at depths of the order of thousands of metres as at the Tahe Field, where not only the intrinsic depth but also variations in the nature and structure of the formation exacerbate the usual challenges in drilling and in casing installation.

Formic acid

Not only formates but also formic acid itself finds application in **drilling fluids**. Its role is in removal of filter cake where this contains large amounts of calcium. Other acids including citric acid and lactic acid have been applied to filter cake removal. In using an acid for filter cake removal it has to be remembered that the brine itself contains inorganic solute and interaction of this with whatever acid is added can affect the pH. Accordingly, in one R&D investigation it was found that, for brines containing formate, acetic acid was the most suitable for filter cake removal. Formic acid is also sometimes used in well stimulation in preference to hydrochloric acid. Formic acid is a weak acid, meaning that its dissociation to form hydrogen ions in water is limited. Whilst on the one hand this might affect its efficacy as a stimulant, on the other hand formic acid is, other things being equal, much less likely than hydrochoric acid to damage steel with which it comes into contact on its way to the stimulation site. It might therefore be possible to organise matters so that, by substituting formic acid for hydrochloric, an acceptable degree of stimulation is achieved with the total avoidance of 'pitting' of steel in the well. Of course, an alternative strategy to acid composition control in the avoidance of steel damage during acid treatment of a well is use of a corrosion inhibitor such as **triethanol amine**. (*See* **Arcasolve™**.)

Fort Yukon AK, CBM at

Slimhole drilling has been applied to coal deposits at Fort Yukon in rural Alaska to assess the viability of **CBM** production from them. The coals there are of lower rank – less mature – than the coals in the more southern 48 states that provide CBM. The locality, already described in this entry as 'rural', is so much so that it is not on the electricity grid and diesel is used to produce electricity. The long-term intention of the CBM investigations is the provision of a means of thermal generation of electricity. Investigative drilling was at depths in the range 390 m to 700 m. An important quantity in CBM viability is the **gas saturation level**. To determine this a sample of the coal is simply allowed to lose its methane content by desorption. This leads to a quantity accordingly called the desorption, a.k.a. total gas measured content, units m^3 of gas per tonne of coal. A sample of the desorbed coal is then placed in an atmosphere of methane and the uptake followed. The amount of methane so taken up on the basis that a monolayer of methane forms in the coal's pore structure, calculated from the Langmuir model, is the adsorption, units m^3 per tonne of coal. This experiment will be carried out at a pressure much higher than atmospheric which, it is intended, will represent that in a coal seam. The desorption and adsorption experiments obviously need to be at the same temperature and 32 °C is a common choice. The gas saturation level is then:

$$1 - [(\text{adsorption} - \text{desorption})/\text{adsorption})] \times 100\%$$

Figures for adsorption and desorption for the Alaskan coals under discussion in this entry are not given in the report which has been drawn on, although gas saturation levels are and will be discussed. The adsorption and desorption figures in the sample calculation below, taken from an article in the research literature for 2009, are actually for a New Zealand coal.

$$\text{desorption } 2.50\ m^3\ t^{-1}\ (32\ ^\circ C)\ \text{adsorption } 3.26\ m^3\ t^{-1}\ (32\ ^\circ C,\ 4\ MPa)$$

$$\downarrow$$

$$\text{gas saturation level} = 1 - [(3.26 - 2.50)/3.26] \times 100\% = 77\%$$

The gas saturation level values for the NZ coal of which the above is one example actually range from < 15% to > 120%. Significant spread was also shown when the Fort Yukon coals were tested in the same way. It is reported that the average was 20% to 30% and that in one canister it was 50%. One can speculate on the physical meaning of variations in the gas saturation level. One is that the methane in the coal and that in pockets of gas are not in phase equilibrium. Another is that the adsorption goes beyond a monolayer, in which case adsorption measurements using Brunauer, Emmet and Teller (BET) instead of Langmuir would be more informative. To apply BET instead of Langmuir would involve no extra experimental work and hardly any extra calculation work. It *might* be that workers in this field observing the enormous variations typified by the NZ coal consider that the refinement of using BET would be of no real meaning. Yet another explanation of the difference is that imbibition as well as adsorption is involved. The values obtained for the Fort Yukon coal were lower than hoped for and are seen as signifying the need to drill more wells than were originally planned. Note that the figures in the box for adsorption and desorption correspond to about 100 cubic feet per tonne and this can be compared with the value given for **Airth** and that for **Bruna**.

Forties pipeline

This originates at one of the platforms in the Forties Field – Forties Charlie – and via another such platform ('Forties Unity') carries oil to Cruden Bay on the coast of north-east Scotland. This is a distance of 169 km. From there it continues on land, there being **pumping stations** along this part of the pipeline, which is 209 km in length. At the place where the pipeline comes onshore, NGL, separated from gas having been brought from North Sea fields to St. Fergus also on the NE Scotland coast, is incorporated with the crude oil. This goes to the oil terminal at Kinneil where it undergoes stabilisation prior to further pipelining to the Hound Point terminal in the Firth of Forth for export. Hound Point can accommodate vessels of up to 350 000 tonnes deadweight. A calculation like that performed for the **Jebel Ali Refinery** reveals that the capacity of such a vessel is:

$$(350000 \times 10^3 \text{ kg}/950 \text{ kg m}^{-3})/0.159 \text{ m}^3 \text{ bbl}^{-1} = 2 \text{ million barrels approx.}$$

This seems to indicate that tankers in the 'supertanker' league can avail themselves of Hound Point. However, an arbitrary definition of a supertanker, which as far as the author is aware does not have the authority of the IMO[24] or other such body, is a tanker of deadweight 500 000 tonnes, exceeding the size that can be received at Hound Point. This entry is being written one month after the seizure of a Saudi oil tanker by Somali pirates in February 2011. It was reported that the tanker was carrying 'almost

2 million barrels of oil' which puts it in the same size range as that calculated above as the maximum for Hound Point. It was taking oil to the US.

Fos Tonkin (Fos-Sur-Mer) LNG Terminal

In France, receiving LNG from Algeria (quite a short distance). It can produce from LNG received seven billion cubic metres per year of gas, or:

$$(7 \times 10^9 \text{ m}^3 \text{ year}^{-1}/365 \text{ day year}^{-1}) = 20 \text{ million m}^3 \text{ day}^{-1}$$

placing the facility in the 'big league'. The interested reader can confirm by calculation that the figure given in weight terms for the **Map Ta Phut LNG Regasification Terminal** is about the same as this and that the figure for **Dabhol LNG terminal** is about a factor of four smaller. (*See also* **Dapeng LNG terminal**, **Rudong LNG terminal**, **St John**, **New Brunswick**.)

FPSO

Floating production storage and offloading vessel. A state-of-the-art FPSO is the Anasuria, built by Mitsubishi and in service in the North Sea. (It was built as an FPSO, whereas some FPSOs are rebuilds and adaptations of vessels previously in service as tankers, as at the **Peregrino Field**.) Anasuria can receive from from subsea wells up to 850 000 barrels of crude oil for transfer either to tankers or to offshore storage installations. The FPSO can be connected to up to fourteen oil risers from the sea-floor at any one time. Its pumps are such that it can receive oil at up to 150 barrels per minute and release it at up to about 600 barrels per minute. It has facilities for **water injection**. Like the FPSO at **Schiehallion**, it is held in position by chains attached to the sea-floor by piles. An FPSO might be a permanent installation or might equally be relocateable. That at the **Bonga** Field is expected to be there permanently. At **Kizomba** an FPSO and a **tension leg platform** are connected to each other providing in effect a composite facility. The very first FPSO operation was in 1977 at the Castellon Field offshore eastern Spain, in a water depth of 116 m. The operator of this FPSO was Shell and its storage capability was < 500 barrels. (*See also* **Buffalo Venture**; **Erha Field**; **Firenze**; **Girrassol**; **Glas Dowr**; **Glitne Field**; **Hæwene Brim**; **Haiyangshiyou 102**; **Jubilee Field**; **Kikeh**; **Munin**; **Norne**; **Northern Endeavour**; **Okoro Field**; **Production rigs, propulsion of**; **Rubicon Offshore, vessels operated by**; **Shelley Field**; **Su Tu Den**; **Uisge Gorm**)

FPU

Floating production unit. (*See also* **Balder**; **Janice Field**; **Troll Field**; **Yttergryta Field**.)

Frac-Pack

When a gravel pack is used to remove sand from crude oil during its passage from reservoir to well some resistance to flow results, with obvious effects on well production performance. That is why a gravel arrangement with a good 'conductivity' will be used. An alternative is hydraulic fracture with gravel pack filtration. The conductivity of the **proppant** will then influence flow, and coating of proppant particles with resin can improve the conductivity. This Frac-Pack approach is adopted widely in many places including the Gulf of Mexico, for example at the **Tahiti Field**. (*See* **Ceiba Field**.)

Fracture pressure

The pressure needed to create the required effect in hydraulic fracture. Once the fracture is created the importance of the fracture pressure is that it is an upper bound on the closure pressure which applies to a **proppant** which occupies the fracture. This in turn affects the conductivity of the proppant. In fact the closure stress is 80% to 90% of the fracture pressure and therefore has units of pressure. The data below relate to a **Brady® proppant** of particular particle size range. The units of conductivity are millidarcy-feet (md-ft), being the product of the permeability (md) and the distance interval over which it applies (ft). These are the usual units not only in oil well matters but also in hydraulic engineering generally. Following the values for the Brady® proppant are those for a resin coated proppant made by Borovichi in Russia[25].

Brady® proppant

Closure stress/p.s.i.	Conductivity at 150 °F/md-ft
2000	23944
4000	5522
6000	1912
8000	351

Borovichi resin coated proppant

Closure stress/p.s.i.	Conductivity/md-ft (undisclosed temperature)
2000	45945
4000	37350

Closure stress/p.s.i.	Conductivity/md-ft (undisclosed temperature)
6000	24560
8000	12099
10000	6188

With either of the proppants for which data are presented there is a strong negative correlation between the pressure needed to create the fracture and the conductivity of the proppant. Once a proppant has been used in hydraulic fracture there is invariably some remaining, not having settled at the site of the fracture. If this is not removed it will of course contaminate products from the well. Removal is called post-stimulation clean-up. Where hydraulic fracture with proppant installation takes place it might be such that its orientation is not the most suitable for the existing **completion**, for example the positioning of the screens and perforations. This can be a factor in the efficacy of well stimulation by hydraulic fracture. (*See also* **Bond™Lite**; **Nitrogen, as a carrier for proppants**; **North Sea, first hydraulic fracture operation in**; **Porosity**; **Tight gas, South American sources of**; **Vernon Field**.)

Freeport TX

The scene of an LNG terminal since 2008, capable of producing from LNG 56 million cubic metres per day of gas. By a calculation similar to that given in detail for the **Hazira LNG terminal** this becomes 24 GW in heat supply terms. The imminent closure of the **Kenai LNG plant** will as noted mean that there is no export at all of LNG from the US. For the Freeport facility to be doubled up so as to include a liquefaction plant would redress this. Proposals by the Australian based Macquarie Group, through its US representation, are under way for such a liquefaction plant and the capacity intended is around forty million cubic metres of gas per day for liquefaction, with export by means of four LNG trains. Commencement of production in 2015 is hoped for. (*See also* **Panama canal, passage of LNG carriers through**; **Shale natural gas**.)

Frigg Field

In the North Sea, straddling the UK and Norwegian sectors, and the scene of natural gas dehydration using **TEG**. (*See* **Vesterled pipeline**.)

Froy Field

Oil field in the Norwegian sector of the North Sea, water depth 120 m. It was productive, with Elf as operator, from 1995 to 2001 and is now being equipped for production to resume in 2014. Redevelopment plans involve a new-build **jack-up** structure with drilling and production capability. The modest water depth is suited to this approach, as it is to transfer of oil from the production facility to a storage vessel on the seabed and from there to **shuttle tankers**. This makes for a very compact arrangement. There will even so be capacity to spare, with the storage and offloading facilities at Froy sometimes receiving oil from other fields.

FSO

Floating storage and offloading unit, to which oil from an **FPSO** can be transferred by means of an **OOL** to await transfer ashore. An FSO will often have single point mooring (SPM), for example that having recently entered service at **Orkid Field**. Note that SPM does not mean a single mooring line: it means that all of the mooring lines converge at a single point on the FPSO, FSO or whatever is being 'moored', the topside of which has to be configured to be suitable for SPM. An FPSO might well have a helipad. Now that double hull tankers are becoming the norm, many single hull ones have been used as FSOs after suitable conversion. (See also **Apsara Field**; **Cepu Field**; **Hanze F2A Field**; **Haiyangshiyou 102**; **Kittiwake**; **Kizomba**; **Kome Kribi 1**; **Molikpaq**; **MOPU**; **Nam Con Son pipeline**; **Okha FSO**; **Okwok Field**; **Palanca**; **Pathumabaha**; **Platong II Project**.)

Fulmar Field

Oil field in the North Sea, having a **pay** of 175 m. The 'Fulmar formation' is composed of sandstone with porosities in the 20% to 30% range and permeabilities in the 500 millidarcy to 4000 millidarcy range (i.e., up to the very high value of 4 darcy). Fulmar is the scene of **STL**. The disconnectability feature of submerged turret systems is central to the functioning of the Fulmar Field facility where a number of **shuttle tankers** receive crude oil in turn from risers supported by the submerged buoy. The buoy is held at the sea-floor by catenary means at a depth of 80 m. In general, **suction anchors** can be used with submerged turret arrangements.

Fusion Bonded Epoxy (FBE)

A substance with good anti-corrosion characteristics used in the insulation of subsea pipelines. Several of the **pipe laying vessels** described in this volume have a coating as well as a welding facility. For example, **Solitaire** has four 'coating stations' and **Audacia** is also equipped for pipe coating and, in general, space will be allocated for coating on such a vessel. Even so, coating might have taken place by the time the pipe, in section or reel form, is taken on board a pipe laying vessel. Whether on- or offshore, with what are the pipes coated, and why? Avoidance of corrosion is one obvious reason for applying a coating, which might be epoxy based, as with the FBE product manufactured by OJS in Houston. The epoxy substance in powder form is applied at 230 °C to the pipe, the outside of which has been cleaned by grit blasting. The 'application system' which OJS use enables pipeline coated with FBE to be laid at a rate of 9.3 km (5.8 miles) per day. OJS also have a process whereby a pipe for subsea use can be coated with polyethylene to form a 'heat shrink sleeve', also for corrosion protection.

FX™

Fixed, as opposed to roller cone, drill bit from Halliburton using **PDC** and currently being introduced where roller cone bits had previously been used. It owes its good performance to enhanced bit life resulting from **secondary cutters** also made of PDC.

G

Gajah Baru Field

This gas field offshore Indonesia is entering production at the time of going to press. The operator is UK's Premier Oil. **Spudding in** of the first appraisal well was in September 2004 and drilling was to a depth of 2746m. Target production is 100 million cubic metres per day for pipeline export to Singapore.

Galoc Field

Offshore the Philippines, discovered in 1981, producing since 2008 and under expansion. Water depths are in the range 290 m to 400 m and the formation is turbidite sand. The **pay** is 2100 m below the seabed. Two horizontal production wells were drilled in 2006, each of which has 1600 m of reservoir contact[26]. (*See also* **Papyrus-1X**; **Rubicon Offshore, vessels operated by**.)

Gamma (γ) rays, detection of in MWD

As well drilling takes its course the γ-ray activity of the rock having been drilled will vary. Such variation provides information on the nature and composition of the geological formation and this is part of **LWD**, which is a term complementary to **MWD**. The unit of γ-ray release for LWD/MWD is the API unit. This has as its basis a particular calibration standard, retained at the University of Houston, the γ radiation emission of which is assigned the value of 200 API units. 200 API units is about twice the value expected for typical sedimentary rock. (*See also* **GPT**; **Potassium, Thorium and Uranium in well formations**.)

Gannet Field

In the UK sector of the North Sea. Ownership is 50–50 Shell and Esso. It is divided into 'satellites', Gannet A to Gannet G. A **fixed platform** at Gannet A receives oil from all of the divisions. This stands in 95 m of water and its supporting legs are held in position by piles that penetrate the sea-floor to an extent of 85 m. Gannet A has a permeability of 0.5 darcy to 2 darcy and a **porosity** >30% and there is **aquifer water drive**. Gas reinjection has been the practice at the field, with resulting development of **gas cap**, removal ('blowdown') of which has been proposed.

Gas cap

A layer of the formation above that containing the oil, where the pores are filled with gas. It occurs naturally through saturation of the oil with gas. Such a 'cap', initially absent but having developed during production as result of of oil and gas reinjection, can be removed by blowdown, and this might be necessary before well intervention is attempted. (*See also* **Captain Field**; **Depletion drive**; **Gannet Field**; **Harding Field**; **Jubilee Field**; **Oseberg Field**; **Rosneft, selected major upstream assets of**; **Sognefjord**, **Yibal Field**.)

Gas, conventions in the pricing of

It was fully explained in the first of this series of dictionaries that, in contrast to solid and liquid fuels for which calorific value prices are based on quantity, e.g., $US per barrel for crude oil, gas prices are established on a heat basis, usually $US per million BTU. The most widely cited benchmark price, in those units, is the Henry Hub price. Perhaps because of the new role of 'unconventional gas', pricing of natural gas on a volume basis – per thousand cubic feet – is becoming noticeable. The web site:

http://www.eia.doe.gov/dnav/ng/hist/n9190us3a.htm

reports a price of $US4.16 per thousand cubic feet for 'the 2010s' as far as they have elapsed. The Henry Hub price on 20 April 2011 is $US4.33 per million BTU. These can be examined for consistency.

> Assigning the gas to which the web site figure relates a calorific value of 37 MJ m^{-3}, one million BTU would be produced by:
>
> $\{[10^6 \text{ BTU} \times 252 \text{ cal BTU}^{-1} \times 4.2 \text{ J cal}^{-1}/(37 \times 10^6) \text{ J m}^{-3}] \text{ m}^3/0.028 \text{ m}^3 \text{ ft}^{-3}\} \text{ ft}^3 = 1022 \text{ ft}^3$
>
> So the price per thousand million BTU would be $US(4.16 × 1022/1000) = $US4.24 which is about 2% lower than the Henry Hub price, and consistency is confirmed.

Gas saturation level

(*See* **Fort Yukon AK, CBM at**)

GasDry

Dehydrating agent for natural gas containing calcium chloride instead of silica gel or a glycol. It is manufactured by Van Gas in Lake City PA. It is capable of bringing natural gas down to the pipeline level of about 7 lb per million cubic feet of gas. A similar product is Peladow DG from Dow Chemicals. Trends are for onshore use of calcium chloride as a desiccant for natural gas to be favoured for modestly productive fields, say up to 0.1 million cubic metres per day which might of course be associated gas. Offshore, the fact that calcium chloride, unlike **TEG**, is non-flammable and non-volatile adds to its attractiveness as does the fact that a calcium chloride drying unit can be designed to be compact.

GaugePro XPR™

Reaming tool from Baker Hughes. When inserted into a well bore it is in retracted state and can be brought into operation as and where required in the well bore. The retraction facility also makes for straightforward withdrawal after use. It commonly uses a **PDC** cutting tool which is also a Baker Hughes product. (*See* **EZReam™**.)

GBS

Gravity base structure[27]. (*See also* **Brent Charlie**; **Hanze F2A Field**; **Jeanne d'Arc Basin**; **Malampaya Field**; **Molikpaq**; **North Adriatic LNG Terminal**; **Oseberg Field**; **Ravenspurn North**; **Statfjord Field**; **Troll Field**; **Yolla-A platform**.)

GDF Suez Neptune

The first **SRV** – a floating facility to return LNG to natural gas ('regasification') – yet, having been commissioned as recently as 2010, and still state-of-the-art. It was previously an LNG tanker and was converted to its present role by fitting three skid mounted regasification units. Each such unit operates in a multi-step fashion as follows. LNG at −160 °C is heat exchanged with propane gas, which is admitted under equilibrium conditions at a pressure such that its temperature is around 0 °C. The LNG evaporates, leaving the exchanger at about −10 °C. Propane condenses but, being present as a single pure compound, does not by the phase rule change its temperature until condensation is 100%. Heat from seawater is then used for two purposes: to further heat the natural gas to ambient temperatures and to return the propane to the gaseous state for another heat exchange passage. The LNG/propane heat exchange is shown in the conventional way – temperature of each fluid against distance along the exchanger, symbol x – in the figure below. This treats the process as though regasification were instant upon exchanger entry because it is at its boiling point and therefore heat exchange is between the gas only and propane. There are things to be said for and against such a view. As noted, the propane does not (*can*not) change temperature as long as there are two phases present in equilibrium.

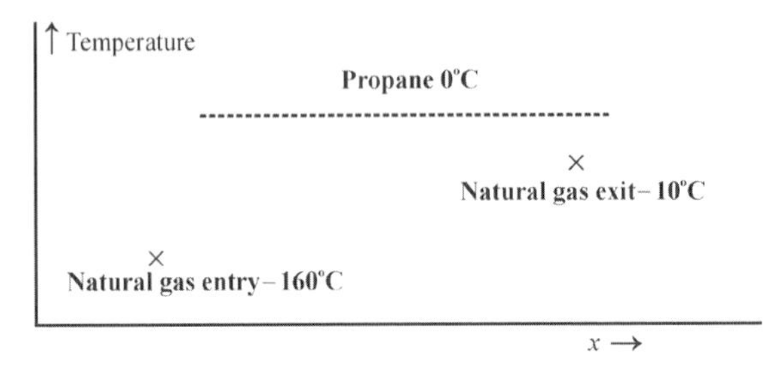

The log mean temperature difference (LMTD) is:

{[0 - (-160)] - [0 - (-10]}/ln(160/10) °C = 54 °C

Suez Neptune, which is capable of producing from LNG 0.6 million cubic metres of natural gas per day, is currently berthed off Gloucester MA. (*See also* **Bahia de Bizkaia Regasification Plant**; **Mizushima LNG import terminal**; **Bridgeport CT**.)

GDU

Gas dehydration unit. (*See* **Glycol dehydration**.)

Gel pig

In such a device a solid gel is used instead of a metallic and/or polymer structure in pipeline maintenance. An obvious point is that the inner diameter of the canister in which a gel pig is supplied will need to be the same as that of the pipeline which it enters, and manufacturers supply accordingly. The gel once made has only a limited shelf life before use, typically four weeks (although for GGPP – see below – it is as high as six months). The gel pig is used in applications including cleaning and debris removal. It might break down totally in use, otherwise what is left of it can be carried out by the production stream. Care must be exercised in the use of gel pigs in natural gas pipelines where hydrate formation is a possibility. One of many examples of gel materials is GGPP – glycol gel pipeline pig – made by Alchemy Oil Field Service whose HQ are in the Aberdeen area. It can be carried along a pipeline for cleaning and debris removal by water, by gas or a by a simple organic agent such as methanol. Another example is the MEG Gel Pig from Aubin Subsea, which has been used extensively in the North Sea and consists of gelatinised **MEG**. A gel pig and a conventional pig (a.k.a. a mechanical pig) can be used concurrently. When a gel pig is used alone there is the obvious advantage that it cannot become stuck in the pipe as mechanical pigs sometimes do! (*See also* **L-Gel**; **SureGL™**.)

Genesis D™

PDC drill bit from Baker Hughes. It is suitable for orientation in directional drilling by a **downhole motor**.

Genesis Field

Oil field in the Gulf of Mexico. The production facility, which is in 792 m of water, is a **spar platform** also suitable for drilling. Mooring is at 14 points on the structure. Mooring lines are composite, comprising chain, mooring wire and then more chain of a different specification than the first segment to a total length of 1340 m. Attachment to the sea-floor is by mooring piles of 8 feet diameter. The platform can receive oil from up to 20 production wells.

Girrasol

Offshore Angola, scene of operation of what is said to be the largest **FPSO** in the world. The hull construction is such that it can remain in situ for up to twenty years without return to dry dock for maintenance. The FPSO has **spread mooring**, sixteen mooring lines emanating as four groups of four at the vessel in a symmetrical arrangement. Each mooring line is attached to the sea-floor by a single **suction anchor**. There are wells for gas reinjection as well as for production. The riser tower concept was largely tried out at Girrassol where there are three such devices in service. Each riser tower supports a bundle of risers and is itself stabilised by attachment to a suction anchor. It has facilities for **glycol dehydration**. (*See* **Rosa Field**.)

Gjøa Field

Condensate field offshore Norway, discovered in 1988 but producing only since 2010. The participants include Statoil and Gaz de France, also GDF Suez who will be production operator. The water depth is 360 m. The production facility is a **semi-submersible rig**, having chains and mooring lines. It is supplied with electricity from by an 'umbilical cable' from onshore. Gas from the field goes to St. Fergus in Scotland and liquids, like those from **Kollsnes**, are taken to Mongstad for refining. (*See* **Norne**.)

Gladstone LNG project

Like **Curtis island**, Gladstone Queensland will become the scene of LNG production from **CBM** (a.k.a. CSG). Commencement of production is hoped for in 2015, participating firms being Santos (Australia), PETRONAS (Malaysia), Total (France) and KOGAS (South Korea). Initial production will be 7.8 million tonnes of LNG per annum from two LNG trains. Malaysia is at the present time, on its own territory, the world's largest producer of LNG having 'overtaken' Indonesia. South Korea is a major importer of LNG and will need to have new sources to cover for that previously obtained from the **Kenai LNG plant**. (*See* **MER-14X**.)

Glamis field

In the UK sector of the North Sea, commencing production in 1989. The formation is sandstone and permeabilities are high – about 1.5 darcy – with an average **porosity** of 15%. There were initially two vertical production wells, drilled so as to accommodate later sidetracking. Oil exiting the wells at the field is taken by flow line to the **FPU** Balmoral. (*See* **Burghley Field**.)

Glas Dowr

FPSO having recently been in service at the Sable Field offshore South Africa, where it receives hydrocarbon from four production wells. There is one gas reinjection well and one well for **water injection**. The water depth at the Sable Field is 102 m and there is single point mooring of the Glas Dowr. A calculation apropos its capacity follows in the box below.

The specifications for Glas Dowr from the manufacturer and available online give a deadweight of 105 000 tons. The load is divided as follows:

Crude oil at full load 657 000 bbl
Slop tanks contents 41 800 bbl
Diesel fuel 16 700 bbl

Total 715 500 bbl

Using a value of 900 kg m^{-3} for the density:

$$\text{Weight} = 715\,500 \text{ bbl} \times 0.159 \text{ m}^3 \text{ bbl}^{-1} \times 900 \text{ kg m}^{-3} \times 10^{-3} \text{ t kg}^{-1}$$

$$= 102388 \text{ tonnes} = 102388 \times 1.1023115 \text{ US tons} = \underline{112863 \text{ US tons}}$$

or

$$102388 \times 0.984206528 \text{ Imperial tons} = \underline{100770 \text{ Imperial tons}}$$

Either figure has reproduced the deadweight to within a few percent, and such simple checks are helpful in engendering confidence in the interpretation of such specifications. The author was unclear whether in those for Glas Dowr the US ton or the Imperial ton was being used.

(*See also* **Munin**; **Pioneer Knutsen**; **Torm Ugland**.)

Glitne Field

Oil field offshore Norway, with some associated gas. Statoil have a major interest in it, along with other companies including Total. The **FPSO** Petrojarl I owned by Teekay is in position there. There are five production wells plus one for **water injection** and one for gas reinjection. The water depth is 110 m. Petrojarl I has **turret mooring** with eight chains each linked to a 15 ton anchor on the sea-floor. The oil production capacity is up to 45 000 barrels per day. There is flaring capacity sufficient for all of the gas in the event that reinjection is not carried out. The **shuttle tanker** Petroskald receives oil from Petrojarl I for transfer to port. The Glitne Field is seen as being suitable for expansion, and formal approval for the drilling of further exploration wells has been granted. (*See also* **Foinaven Field**; **Gjøa Field**.)

Glomar Explorer

Drill ship, a conversion of a vessel initially built for a particular mission by the US Navy in which the CIA were also involved. Activity since the conversion has included drilling at the **Agbami Field**. It can operate in up to 3050 m of water and can drill to 9000 m. It has **dynamic positioning** and can move under its own propulsion at 10 knots.

Glutaraldehyde

Structural formula: $(CHO)–(CH_2)_3–(CHO)$. (*See also* **Bio-Clear™ 242D**; **DBNPA**.)

Glycol dehydration

A widely used means of removing water from natural gas. The 'glycol' might be diethanol glycol or **TEG**. A typical target dryness will often be 7 lb of water per million cubic feet of gas. A related calculation follows.

10^6 ft^3 of gas under 'standard' conditions contains 1×10^6 mol.

7 lb of water contains $(7/2.205) \times 0.018$ mol = 0.057 mol

Level of water in the natural gas after treatment with glycol =

$[0.057/(1 \times 10^6)] \times 10^9$ parts per billion = 57 p.p.b. (0.057 p.p.m.) to the nearest whole number.

There are glycol dehydration units that will take the moisture content of the natural gas half an order of magnitude lower still. In terms just of the quality of the gas this might seem like a needlessly exacting degree of moisture removal. The reason the gas is so thoroughly dried is, however, that the temperature of subsea pipelines is low enough for the dew point to be encountered with higher initial levels of water vapour in the gas, leading to condensation of water in the pipeline. A further calculation follows.

> Imagine natural gas in a pipeline at 100 bar. If the water level were 10 parts per million (molar basis) the pressure of water would be:
>
> $$100 \times 10^{-5}\ \text{bar} = 10^{-3}\ \text{bar}$$
>
> Now from steam tables the saturated vapour pressure of water at 0.01 °C is 0.006 bar, and it is clear without refinement of the calculation that if water in natural gas in a subsea pipeline – or an above-ground one in cold conditions – were in the region of 10 p.p.m. it would be at pressures of the order of its equilibrium vapour pressure and so there would be condensation to the promotion of the formation of hydrates as well as pipe corrosion. That is the reason for the very stringent drying.

Note that the specification given and applied in this entry is for natural gas destined for pipeline transfer. If it is intended instead to convert it to liquefied natural gas (LNG) a different specification applies. (*See also* **Bahrain Field**; **Gel pig**; **MEG**; **Molecular sieve, use of in natural gas dehydration**; **Platong II Project**; **Silica gel, use of in natural gas dehydration**.)

GMS Endurance

Newly commissioned **jack-up** vessel, built in Abu Dhabi. It can operate in water depths of up to 65 m. Another new jack-up vessel from GMS (Gulf Marine Services) is the GMS Endeavour which has many specifications in common with the Endurance. Like many of the most recent jack-up vessels, these are intended for offshore wind turbine installation as well as for drilling for oil and gas. (*See also* **Offshore Freedom**; **West Epsilon**.)

GO FLOW™ 126

Drag reducing agent for use in crude oil pipelines from Weatherford International. It is used at 5 p.p.m.to 50 p.p.m. Its sister product GO FLOW™ 125 is for pipeline flow of refined material such as gasoline and is used at 3 p.p.m. to 30 p.p.m. Each is a polymer.

Golden Pass LNG terminal

Close to Port Arthur TX, having come into service in March 2011. It receives

LNG from Qatar. When the terminal is operating at nameplate capacity it will evaporate 15.6 million tonnes per year of LNG. There is also the Golden Pass pipeline, taking the gas once 'regasified' to other pipelines for intra- and interstate distribution. It is 75 km in length and has been judged not to need a **compressor station**. (*See* **Sabine Pass**.)

GPT

Gravel pack tool. The role of a gravel pack at a well **completion** in removing sand thereby preventing erosion has been explained in the coverage of completion. Condition monitoring of a gravel pack is helpful and is possible by means of a GPT. Such a device from Sondex uses γ-rays as, of course, does one aspect of **MWD**. The GPT releases γ-rays from a caesium-137 source and the extent to which the γ-rays *return* to the tool is the basis of assessment of gravity pack condition. Low return indicates voidage filled with liquid whereas high return indicates unfilled voids or even absence of gravel. The Sondex device can operate at high temperatures and pressures.

Grain LNG Terminal

On the Isle of Grain in the Medway River, Kent, England. Originally only for storage of LNG, it was extended in the early 2000s to become a terminal for import of LNG and regasification, for which it is set up with **SCV**. It receives LNG from diverse sources including Trinidad, Algeria, Egypt and Qatar. In early 2011, Centrica, operator of the Grain Island facility, entered into an agreement with Qatargas whereby 2.4 million tonnes per year of LNG will be delivered to Isle of Grain. There it will of course be returned to gas and made available to domestic and commercial users. It should perhaps be noted that this gas from Qatar is for the UK 'supply-at-large' and not for any particular enterprise such as an electricity utility. (*See also* **Fos Tonkin (Fos-Sur-Mer) LNG Terminal**; **North Adriatic LNG Terminal**; **Perlite**.)

Gravel pack carrying fluid

(*See* **ClearPAC**, **HydropacSM**.)

Gravel pack screen

An important part of the set-up for **completion**, some such devices comprise simply a piece of carbon steel tube such as might be used as drill pipe ('base pipe') into the curved surface of which a suitable number of holes is drilled. 'Suitable' is not being used in a vague sense: the fluid entry per foot of pipe is a factor in performance and needs consideration for particular wells. The drilling is of course done before admittance to the well and is a totally straightforward operation. This contrasts with perforation which applies to the well casing and involves in situ detonation.

Greater Green River WY

Highly productive gas field having more than once undergone hydraulic fracture on account of the low permeability, 3 millidarcy to 5 millidarcy (i.e., it is tight gas). The more recent hydraulic fracture has involved nitrogen as a carrier for the **proppant**. Also in Wyoming Cave Gulch, there is a tight gas field with a **pay** of 270 m. Information on some other tight gas fields in the USA and Canada is given in the table below.

Name and location	Details
Wilcox/Lobo, S. Texas	Permeability from 1.5 md down to > 0.1 md. Water produced with the gas at a rate of about a barrel per million cubic metres of gas.
Wattenburg Field, NE Colorado	Over 4000 wells taken to the stage of **completion**. Extensive application of hydraulic fracturing.
Vicksburg formation, TX.	Well casing collapse experienced initially. Remedies included smaller diameter casing and a greater quantity of cement per length of casing installed.
Dunvegan, Alberta	Permeability < 1 md. Hydraulic fracture with a non-aqueous fluid.

The *total* reserves of tight gas in the USA and Canada are believed to be of the order of 20 trillion cubic metres. (*See* **Piceance Basin**.)

Gryphon

FPSO in service in the North Sea which, in February 2011, experienced storm damage. The FPSO has catenary mooring, and four of its ten anchor chains broke as a result of the storm. There was however no leakage of oil. Two tugs later secured the Gryphon to enable the broken anchors to be replaced. This entry is being written within a week of the incident, and the Gryphon is securely in position but will not be used for production until an independent enquiry has been completed. Its operators had factored into their expected 2011 production figures 18 400 barrels per day from Gryphon. (*See also* **Culzean discovery**; **McClure Field**.)

GSF Adriatic VI

Jack-up rig owned by Transocean having recently seen service in Nigerian waters. It can operate at water depths up to 14 m and is supported on

spud cans of diameter 46 feet. Its **drawworks** are rated at 2000 h.p. Its **POB** is 108. (*See* **Okoro Field**.)

GSF Britannia

Jack-up vessel owned by Transocean having entered service in 1968 although, obviously, there has been retrofitting of facilities and devices over that time. Its current status is 'cold stacked'; it is berthed in the expectation of future use and does not in the meantime have a crew. Its **drawworks** are rated at 2000 h.p. and a composite **blowout preventer** incorporates both ram and annular types. It can drill in up to 200 m of water. Also currently in cold stacked status (in Ghana) is its sister **jack-up** rig GSF High Island IX. It recently served at the **Okwok Field**.

GSF Celtic Sea

Semi-submersible rig from the Transocean fleet. It can operate in up to 1750 m of water and its **derrick** can take a load of 700 tonnes. It is moored by wire rope and anchor chains (i.e., it does not have **dynamic positioning**). Its **drawworks** are rated at 3000 h.p. It has recently seen service at the **Appaloosa Field**. (*See also* **Amirante**, **Transocean Arctic**.)

GSF Grand Banks

Semi-submersible rig which in November 2009 received repairs by SFS Overlay whilst working offshore Newfoundland.

GSF Jack Ryan

Drill ship having recently seen service in Nigerian waters. Built in 2000, it can drill in up to 3050 m of water and has **dynamic positioning**. There was an accident on board the vessel in July 2010 when three crew members were thrown overboard; two of them were rescued.

GSF Manhattan

(*See* **Guendalina Field**.)

GSP Bigfoot 1

Vessel with pipe laying capability also with lifting capability having entered service in 2010. Its operator GSP is headquartered in Constanta, Romania. In its pipe laying role Bigfoot 1 uses **S-Lay**. Bigfoot 2, from the same 'family', is a **semi-submersible rig** of some variability of application, having been used in drilling and in production. Bigfoot 3 is an accommodation vessel, providing safety and comfort for those involved in such tasks as pipeline laying and platform maintenance. Bigfoot 1 and 3 are both of recent construction, having been built respectively in 2010 and in 2009. By marked contrast Bigfoot 2 initially entered service almost 35 years ago, and now has rebuild status. The first assignment of Bigfoot 1 was in the Black Sea in laying pipe at the Akcakoca gas field. She was accompanied by the support vessel GSP Vega, also by another support vessel called Amber II. Accommodation for those involved with the pipe laying at Akcakoca is provided, appropriately, by Bigfoot 3. According to the latest information in the public domain at the time of going to press, Bigfoot 2 is in port at Constanta.

GTN

Gas Transmission Northwest, a pipeline for natural gas with its origin at the border between British Columbia and Idaho (which is also, of course, an international border) and terminating at the Oregon–California border. It has a throughput of 815 cubic metres per day and has thirteen **compression stations** with a combined h.p. of 516 200. They are placed 50 to 60 miles apart. Depending on gas prices and on the season, something like 70% is finally used in California. (*See* **Alliance Pipeline**.)

Guendalina Field

Gas field offshore Italy in the Adriatic Sea, at the time of writing approaching commencement of production. The operator is Eni. The water depth is 20 m and the reservoir depth is 3000 m. The formation is sandstone with **porosity** 22% to 23%. Drilling of two wells has been carried out by the **jack-up** rig **GSF Manhattan**. A platform deck is being set up above the wells and this is for further exploration and drilling after the departure of Manhattan. Initial production from Guendalina will be by tie-back to the existing **Amelia platform** which is owned by Eni. The target production set by Eni for the first year of production at Guendalina is 4 million cubic metres.

Gullfaks

Oil and gas field in the Norwegian sector of the North Sea, water depth 135 m. The operator is Statoil and there are three satellites - Gullfaks South, Rimfaks and Gullveig - which are being developed alongside Gullfaks itself where there are three **fixed platforms**, termed A, B and C. Production from Gullfaks A began in 1986. Gullfaks B and Gullfaks C came into operation in respectively 1988 and 1989. Transport of oil from the platform is by **shuttle tanker**. Associated gas is piped to an onshore terminal close to Stavanger for subsequent distribution in and beyond Norway. Gullfaks A receives for storing and offloading oil from other fields including the **Visund Field**.

Gus Androes

Jack-up rig, part of the Noble fleet. Built in Venezuela and entering service in 1982, it can drill in up to 90 m of water and to well depths of >7500 m. It recently received **UWILD** in the UAE from Lamprell as a precursor to modifications which amongst other things will increase the maximum **POB** to 65.

Gusto CTV

Drill ship, a modification of an established type, built expressly for **coiled tubing drilling** (CTV = Coiled Tubing Vessel). It can operate in water depths up to about 2300 m and has **dynamic positioning**. Always with coiled tubing drilling, space to accommodate the reels is an issue, and it sometimes happens that once a drilling rig is in position for such a purpose a supply vessel has to bring the reels to it periodically. When Gusto CTV was configured, space for the reels was carefully factored in.

Guwahati-Siliguri products pipeline

The first products pipeline to be laid in India, extending to a length of of 435 km. The pipe diameter is 8⅝ inches. It carries refined products from the refinery at Guwahati to three collection points, conveying them in a westerly direction. Initially having a capacity of 0.481 million tonnes (about 4 million barrels) of refined product per annum, by 1982 additional **pumping stations** had raised that to 0.559 million tonnes per annum. (*See* **Botswana, oil products in**.)

GWC

Gas water contact, a fairly self-explanatory term indicating the boundary between natural gas and water in a well. However its numerical value can be a little imprecise, as at reservoir pressures methane and water are not totally immiscible. As an example, at Stanley-1, a well at a natural gas prospect in Papua New Guinea, GWC occurs at 3137.5 m **BKB**. (*See* **Pickerill Field**.)

GyroTracer™

MWD instrument based on the **north seeking gyro**, manufactured in Sweden by Stockholm Precision Tools, having found recent application at the **Piceance Basin**.

Hac Long[28] Field

Gas and condensate field offshore Vietnam, discovered in 2009. Petro-Vietnam and Petronas (Malaysia) will be the joint developers. The field is 100 kilometres from the coast.

Hæwene Brim

FPSO from the Bluewater fleet, though having had a previous owner. It has had a long period of working in the Pierce Field in the North Sea in 85 m of water on behalf of Shell, where it receives 30 000 barrels per day for offloading on to **shuttle tankers**. There are facilities for **water injection**.

Haiyangshiyou 102

FPSO of capacity 39 000 barrels per day. Built in 1990, it has single point mooring. In April 2011 whilst the FPSO was in operation in Bohai Bay the mooring failed during rough weather. There were no injuries and there was no leakage of oil. Haiyangshiyou 102 was subsequently secured and tugged to safety.

HAM 318

Trailing suction dredger, the largest such in the Van Oorden fleet. Its pump rating is 1100 kW, very close to the nameplate pumping capacity of **Orisant**. HAM 318, together with two **cutter suction dredgers** also from the Van Oorden stable, has recently been in service on behalf of the South Oil Company of Iraq. This undertaking has included the widening of a river mouth to accommodate a loading buoy. (*See* **Kochi Port**.)

Hanze F2A Field

In the Dutch sector of the North Sea, sea depth varying between 1340 m and 1478 m. The production (and storage) platform is a **GBS**. Drilling was by the **jack-up** rig **Ensco 100**, a sister vessel (same ownership) of **Ensco 70**. Oil is conveyed by pipeline to an **FSO** with single point mooring and gas to a terminal at Den Helder.

Harding Field

Oil field in the North Sea. Oil production began there in 1995. A **TGP** 500 **jack-up** rig, another one of which is in service with BP in the Caspian Sea, is used in production. **Water injection** using in situ water from an aquifer in the formation has been used to stimulate oil production, a process known as **aquifer water drive**, at the Harding Field. **Electric submersible pumps** have been used to move the water and a production rate of 85 000 barrels per day of crude oil attained. Note that this method eliminates the need for a water injection well. There are measures in place to produce from the **gas cap** at the field and to receive the gas at the existing jack-up platform at Harding. (*See* **Shah Deniz**.)

Hassi Messaoud Field

The largest oil field in Algeria, discovered in 1956. Its productivity has varied over the years and attempted improvement by hydraulic fracture in the 1990s met with less success than was hoped for. Its **pay** is at depths of around 3400 m. Five of the 1000+ wells at the field have recently been subjected to **coiled tubing drilling** which was also underbalanced and for part of the time used an **impregnated drill bit**. An interesting point made in a report of the operation is that the pressure at the site of installation of a sidetrack has to be 'adequate to induce underbalanced conditions'[29]. Current production is 0.375 million barrels per day, putting Hassi Messaoud in the league of 'giants'. (*See* **Bir Seba**.)

Hawkins Field

Extension of the life of this field is proposed by use of nitrogen injection at a cost of $US340 million. An extra 40 million barrels of oil will over a 25-year period be obtained from the field. At $US80 per barrel the payback time will be:

$$\$US340 \times 10^6/[(40 \times 10^6/25)\ \text{bbl year}^{-1} \times \$US80\ \text{bbl}^{-1}] = 2.7\ \text{years}$$

an intuitively reasonable result which might be on the high side through not having taken the associated gas into account. It is, however, counterintuitive that in a calculation in a financial paradigm the associated gas would have a strong effect. In fact at the field the associated gas becomes diluted with the injected nitrogen which has to be 'rejected' before sale. (*See also* **Cantarell Field**; **Nitrogen rejection**.)

Hawtah Field

In Saudi Arabia, producing about 20 000 barrels per day of oil and in decline. **Water injection** has been applied, with the effect of introducing

SRB and hence hydrogen sulphide.

Hazira LNG terminal

One of three such facilities in India, entering service in 2006. Its capacity is 2.5 megatonnes per year, in thermal terms:

$$[2.5 \times 10^9 \text{ kg} \times 55 \times 10^6 \text{ J kg}^{-1}/(365 \times 24 \times 3600)] \text{ W} = 4 \text{ GW}$$

There are both **ORV** and **SCV** facilities. (See **Dabhol LNG terminal**.)

HCR 506ZX

Drill bit from Hughes Christensen, used at the **Z-16 ST** well at Sakhalin where, after sidetracking, a well of measured depth 9788 m and of **TVD** 5588 m was drilled. It is very simple to show by mensuration that the **AHD** would have been:

$$[9788^2 - 5588^2]^{0.5} \text{ m} = 8036 \text{ m}$$

These are of course very large drilling depths. The HCR 506ZX is of diameter 12¼ inches and uses a diamond cutting surface.

Helena

'Discovery' in the North Sea containing gas and condensate. Valiant Petroleum (HQ in the UK) have a major interest and are currently seeking 'strategic partners' in the development, an appraisal well having been drilled in 2011. Viability will depend on there being a means of tying back production facilities to existing infrastructure, possibly at the Magnus Field or the Don Field. Valiant, founded in 2004, are noted[30] for their development of small fields. A number of their other current activities are described in the table below.

Banquo Field	Containing oil and gas, close to Helena and being developed jointly with it.
Causeway Field	Close to Cormorant, recent commencement of oil production.
Katherine prospect	Drilling for oil imminent and production hoped for in 2012. Tie-back to Brent or Ninian planned.
Handcross prospect	Potentially productive of oil, 15 km north-west of the **Foinaven Field**. Production hoped for in 2013 using an **FPSO**.

Bourbon prospect	5 km east of the Eider Field to which it will be tied back for oil production, or if not to Eider to another existing facility.

From the selection above of Valiant's activities it is clear that there is reliance either on tie-back or on FPSO usage, as would be expected for smaller fields. Operators of such fields are more vulnerable than the 'majors' in such matters as taxation at oil fields.

HGA-37

An organic based **hydraulic fracturing fluid**, containing kerosene or possibly diesel as such a base, will require a gelling agent for 'frac' use. HGA-37 is one such from Weatherford and contains a phosphate gelling agent. It is one of a family of products and they differ chiefly in temperature range of application. HGA-37 can be used at up to 121 °C whereas HGA-702 can be used at up to 149 °C. The 'recipe' for HGA-37 is 5 gallons to 10 gallons per 1000 gallons of hydrocarbon base. Where the conditions of hydraulic fracturing fluid production by such means are very cold, particle deposition is possible and a freeze point (distinct from freez*ing* point and analogous to the cloud point of an oil) is important. This features in the specifications of some of the product range. (*See* **WGA-1**.)

Hides Field

Onshore gas field in Papua New Guinea, discovered in 1987 and entering production in 1991, at which stage gas from two production wells was subjected to treatment including **condensate stripping** before being passed along for use in electricity generation. The condensate is distilled to form diesel for the local market and also naphtha. The total production of gas for 2009 was 154 million cubic metres. In the planning for the quinquennium 2009 to 2014 some gas from Hides will be committed to the **PNG LNG project**.

HNBR

Hydrogenated nitrile butadiene rubber. It finds applications in O-rings in well casings, also in the **annular blowout preventer** and the **ram-type blowout preventer**. It can also be used in a **pipeline repair clamp**.

Horseshoe Canyon

In Alberta, scene of the first **CBM** production in Canada. Current production is 14 million cubic metres per day. This is expected to rise by about 40% over the next year as new wells come into use. The reserve is dry, so nitrogen fracturing is preferred to hydraulic fracture. Interestingly, coal layers bearing **pay** of CBM are separated by rock formations containing pay of conventional natural gas, and the two can be produced at the same well. Such dual production would be expected to be accompanied by at least trace amounts of condensate and the coal layers have to be targeted in the nitrogen fractures, as many as 20 'fracs' being common in the wells at Horseshoe Canyon. (*See also* **Curtis Island**; **Virginia, CBM in**.)

Horton Bluff shale natural gas

In Nova Scotia and New Brunswick, being developed by Triangle Petroleum Corporation. Typically, two vertical wells were followed by a horizontal one and there has been hydraulic fracturing. The location is such that existing pipelines with capacity to spare could be used to carry Horton Bluff gas to markets in Canada and N.E. USA.

Huabei Field

Oil field in China, in the Hebei Province. Its current production is typically 87 000 barrels per day, and this is within the capacity of the nearby Huabei refinery. Production of oil at the field was about twice this 25 years ago and higher still in the late 1970s. When in 1978 annual production of domestic oil in China reached one million tonnes that was seen as a landmark, and it would not have been possible without the contribution from Huabei. PetroChina Huabei Oil Field Company, who operate the field, also produce **CBM** from the Qinshi Basin which is within the 'Qinshui basin–Huabei Oilfield unit'. The discovery was made in 2006, since when more than 850 wells have been drilled and 436 million cubic metres of gas have been produced. Concurrent CBM and oil activity also takes place at the **Jatibarang Field**.

Huldra Field

Gas and oil field in the Norwegian sector of the North Sea, producing since late 2001. The water depth is 125 m. There are six production wells, which were drilled by **Maersk Gallant**[31]. Depletion is now evident and production beyond 2014 is not expected.

Hull, of an offshore production platform

The part of the platform structure below that occupied by persons and plant, which will be partly above the water level and partly below it. Its buoyancy is provided by a number of pontoons. At the Neptune Field in the Gulf of Mexico there is a **tension leg platform** operated by a group of companies including BHP Billiton and Marathon Oil. There have been difficulties due to what a spokesperson for BHP Billiton described in a statement as 'anomalies' in the hull, which was fabricated in Port Arthur TX and weighs almost 60 000 tonnes. Personnel were withdrawn from the platform so that repairs and reinforcements to the hull could commence.

Humin P 775

Commercial additive for drilling muds made by treating **leonardite** with potassium hydroxide. It finds application in water based muds and in emulsions. In the former it tends to raise the pH, being itself the salt of a strong base and a weak acid. In the latter it aids the stability of the emulsion by reason of its hydrophilic nature.

Hydra-Jar®

From Smith Services, a Schlumberger company, a range of **drilling jars**. Hydra-Jar® is available in the outer diameter range 3⅜ inches to 9½ inches and can function at temperatures up to 260 °C. Further information is provided in the boxes below.

> The 6½ inch o.d. variant can provide a detent force up to 0.8 MN
>
> Its weight is 1850 lb equivalent to 8.2 kN
>
> The detent force is therefore two orders of magnitude higher than the weight. The detent force is not a gravitational effect. It is in fact due to movement of one part of the drilling jar relative to the other using energy provided by previously introduced strain. The absence of gravity dependence is a factor in the use of the drilling jar in **directional drilling**[32].

This and the similar calculation appertaining to **Sup-R-Jar®** can helpfully be considered together. They indicate that the detent force is one order of magnitude higher than the **WOB** and two orders of magnitude higher than the weight of the drilling jar. Interesting though these figures are they must be seen as no more than, at most, a rough semi-quantitative guide expressible as:

> **drilling jar weight** < detent force
>
> detent force > WOB
>
> These inequalities should be compared with the calculated results for **TT®**.

(See **Yield strength**.)

Hydraulic fracturing fluid

Having been used in oil and gas production since about 1940, most are water based although, because of surface tension effects, a hydrocarbon based fracturing fluid might be more effective where the formation permeability is low. Another reason for preference of an organic base fluid is that where clay is present water might cause it to swell. An aqueous or non-aqueous fracturing fluid will use suspended material, possibly sand, a ceramic, glass beads or aluminium pellets and known as a **proppant**, which will deposit in a fracture created by the fluid and prevent it from closing under the weight of the formation when fluid passage ceases. The permeability of the proppant once in place will be such that oil passage through it is straightforward. Catalysts for chemical processing are often on a ceramic support. 'Spent catalyst disposal' is an important issue. A particular catalyst will not be approved for use unless the safety of its entire life-cycle, from manufacture to disposal, is proven. Obviously 'everybody wins' when, as sometimes happens, spent catalyst is used as a proppant. We note in concluding this entry that it is good practice to incorporate into an aqueous fracture fluid a biocide capable of killing **APB** and **SRB**. The biocide BE-3TM from Halliburton is expressly for aqueous fracturing fluids. (*See also* **ELWC**; **Ensco 100**; **Greater Green River WY**; **HGA-37**; **Piceance Basin**; **San Jaun Basin NM**; **Severo-Varyoganskoe**; **Shale natural gas**; **South Arne Field**; **Stim Star**, **WGA-1**.)

Hydrogen cracking

A.k.a. hydrogen embrittlement, a phenomenon in welding, particularly relevant to pipeline welding. It is due to diffusion of hydrogen from the welding electrodes into the metal, with resulting cracks. (*See* **X80 steel**.)

Hydropac^{SM}

This conveys solids for well **completion** packs by suspending them in an aqueous gel which, on ultimate settlement of the solids, leaves little residue and a pack with good permeability. (It is a Halliburton 'service': note that superscript SM rather than TM applies when a service rather than a product is registered by its developer.) Hydropac^{SM} at the commencement of passage to the well contains 6.8 kg of solid per gallon of gel.

I

IADC

International Association of Drilling Contractors. (*See also* **BT1,2,3,3H**; **Dull bit**; **Matrix body**.)

Icebreaker Sakhalin

Based at **Orlan**, having ice breaking as its primary function but also capable of skimming oil spills and of fire fighting. It entered service in 2005 having been built in Finland. The temperatures it encounters are down to −40 °C. Propulsion is by two Azipods[33] each rated at 6.5 MW. In an Azipod the propeller is driven by electricity which originates at a diesel generator. A bonus from Azipod usage is that the use of passive electrical cables instead of mechanical connectors to provide power to the propeller eliminates vibrations. This is a fairly minor point in comparison with the feature of Icebreaker Sakhalin which makes it unique. When travelling unimpeded by ice it does so in the conventional style of bow first. However, when it is required to break ice it moves *stern* first, presenting a larger cutting edge to the ice than the bow would have done. This is termed by the shipbuilders a ‘double-acting concept’. (*See* **Salinity, of produced water**.)

IFV

Intermediate Fluid-type Vaporiser. In such a device the intermediate fluid, which is sealed in and does not ‘flow’ through the exchanger, is vaporised by seawater. This is then heat exchanged with LNG, condenses back and in releasing its heat of vaporisation evaporates the LNG. The gaseous methane is then raised from its boiling point to ambient temperature by exchange with seawater. So, as with the **ORV**, all of the heat required for LNG regasification is obtained from seawater but, in contrast to an ORV, with the involvement of an intermediate fluid that acts by phase change, which of course seawater never could not least because being two-component it does not have a single boiling point. Propane (boiling point −42 °C) is a possible choice of ‘intermediate fluid’. In addition to their use at the **Shanghai LNG Project**, IFVs are in service at LNG reception facilities in Japan, which is the world’s largest importer of LNG.

Impregnated drill bit

One in which, instead of there being pieces of diamond, diamond particles are mixed with a metallic powder and this mix forms the cutting edge. Diamonds in the mix are of particle size 40 mesh to 50 mesh (about 0.4 mm). (*See* **Carter-Knox Field**.)

IMPU

Injected moulded polyurethane, used to coat pipelines to provide them with thermal insulation, that is, to raise the thermal resistance of the pipe–seawater interface. Such insulation has two benefits. First, it helps keep crude oil above its *cloud point*. Below the cloud point, solid will appear and this makes for difficulty in flow. Secondly, where gas is present in the pipeline, hydrate formation occurs at sufficiently low temperatures, so the insulation helps in preventing this. Time must be allowed for curing of the IMPU coating before installation of the pipe by **J-Lay** or by **S-Lay**. Injected moulded polypropylene (IMPP), the propylene analogue of IMPU, finds similar use, and both are available from OJS in Houston. Both IMPU and IMPP can be applied to risers as well as to pipeline.

Incheon LNG terminal

Entering service in 1996 with the capability to regasify 360 tonnes per hour, and having undergone expansion since to become one of the largest such facilities in the world with a capability of 3870 tonnes per hour. **ORV**s are used, although the fluctuation of seawater temperature with time of year has been a factor in their operation. There are also **SCV**s. (*See* **Qalhat LNG plant**.)

Independence Hub

Region of gas and condensate production in the Gulf of Mexico. Current daily yield is 28 million cubic metres of gas and 5000 barrels of condensate. Unusual in general terms but less surprising in view of the depths (see table below) is the fact that the production platform is a **semi-submersible rig**, which is moored by a proprietary rope material containing some polyester. The production wells currently delivering at Independence Hub have been assigned names of their own and a selection of these is given in the table below.

Well	Details
Atlas	Drilled 2003. 55 m of gross **pay**. Water depth 2743 m. Atlas NW a 'satellite'.

Well	Details
Jubilee	Drilled 2003. 25 m of net pay. Water depth 2682 m.
Merganser	Drilled 2001. Water depth 2048 m.

The above wells do not come near to fully utilising the production capacity of the semi-submersible rig but several other wells are to be developed and will eventually be tied back to it.

INHIBI-SEAM

Basis of **drilling fluid** for **CBM** manufactured by AMC (HQ in Western Australia). Supplied in powder form it gives a solution of pH 7.5 when added to water at a concentration of 1% and is used at concentrations up to 5%. In water it releases chloride ions and a 5% solution is 0.2 molar in chloride. Approved (by AMC) polymers can be incorporated to adjust the viscosity. (*See* **USA, selected CBM reserves in**.)

INNOVERT®

An **invert emulsion** manufactured by Halliburton. Having a base of kerosene and **LTMO**, it contains barite ($BaSO_4$) as a densifying agent. Scenes of recent usage include the Alwyn Field in the North Sea. (*See also* **Vesterled pipeline**; **Frigg Field**.)

Invermul®

Drilling mud ingredient from Halliburton. When added to a kerosene base in amounts up to 34 **ppb** (more for use at higher temperatures) it forms an emulsion. Itself an organic liquid with a flash point of 69 °C, it requires lime as a complementary reagent in an amount half that of the Invermul®. It was recently used in drilling at Unayzah in Saudi Arabia.

Invert emulsion

Oil based **drilling fluid** containing an emulsified aqueous phase (water or brine). The oil is the continuous (external) phase and the aqueous material the internal phase. One such having 'North Sea compliancy' is **INNOVERT®**.

Iran Khazar

Jack-up rig, built in Iran in 1995. It has recently seen service offshore Turkmenistan. It can operate at water depths up to 90 m and can drill to depths greater than 6000 m. (*See* **Astra**, **Dzheitune Field**.)

J-Lay

S-Lay involves there being a point of inflection in the profile of a pipe in its installation and the stress that this causes can be prevented if the profile is a 'J' shape instead, hence the term 'J-Lay'. It tends to be used at greater seabed depths, or where sea conditions are vigorous. Otherwise, procedures are as for S-Lay, with a stinger and 'welding station'. It has been noted that with the installation of risers as opposed to pipelines J-Lay is preferred. (*See also* **Acergy Condory**; **Yttergryta Field** *inter alia*.)

Jack/St. Malo Development

In the Gulf of Mexico, production expected to commence in 2014. The fields, even though they are being developed jointly, are of very different sea depths: 2134 m for Jack and 640 m for St. Malo. They are separated in distance by 40 km. Chevron, Maersk and Statoil each have an interest. Drilling of the production wells is expected to be from the drill ship *Discoverer Clear Leader*, a Transocean vessel as is **Discoverer Enterprise** which belongs to an earlier generation of drill ships. At the time of writing this entry Discoverer Clear Leader is already in position in the Gulf. It can operate at water depths up to 3660 m, so can cope with the conditions, deep though they certainly are, at Jack. It can drill wells 12 000 m deep and has a **POB** of 200. Production at commencement of Jack/St. Malo is expected to be 170 000 barrels per day. The oil, once produced at Jack/St Malo, will be taken 136 miles by pipeline to a platform operated by Shell.

Jack-up

Drilling vessel whereby, once the vessel is over the site to be drilled, legs are put down to the sea floor. After the vessel has been so stabilised it is raised by a jack to a suitable height above the sea for drilling. What was previously on the deck of the vessel becomes in effect the topside of the drilling structure. Such rigs are limited to modest sea depths. At, say, the time that North Sea oil production began such sea depths were about the limit, and this type of drilling vessel was widely used. Many of these have remained in use for work in shallow water with the result that there are jack-up vessels that have been in service for 30+ years, e.g. **GSF Britannia**. There are however recent new-build jack-up rigs, e.g. **Al Ghallan**, for scenes of drilling where the water depth is of the order of 100 m. A jack-up rig needing to be moved from one scene of drilling to another nearby can be set up to be skidded. This was so at the Dzheitune Field. (*See also* **Astra**; **Atwood Aurora**; **David Tinsley**; **Deep Panuke Field**; **Ensco 100**; **Iran Khazar, Jasmine Field**; **Khvalynskoye Field**; **Lloyd Noble**; **Maersk Gallant**; **Maersk Guardian**; **Mat support**; **Nang Nuan Field**; **Nini and Cecilie fields**; **Noble Al White**; **Offshore Freedom**; **Okoro Field**; **Papyrus-1X**; **Rashid Field**; **Rowan Gorilla III**; **Sapele well**; **Shah Deniz**; **Shallow seawater, possible difficulties with**; **Spud can**; **Stella development**; **TGP**; **Thebaud platform**; **Noble Tommy Craighead**; **Tristan North West**, **Troy Williams**; **West Epsilon**.)

Janice Field

In the North Sea, in 80 m of water and only 12 km from the Fulmar Field. Since production began in 1999 it has yielded up to 30 000 barrels per day with associated gas. Oil from production wells at the Janice Field is taken to an **FPU** called Janica and there are also wells for **water injection**. Janica, has a **POB** of about 30, and is an adaptation of a **semi-submersible rig** previously called 'West Royal'. Some of the associated gas from Janice is used to generate power for the operations: the remainder is pipelined to the Judy platform and combined with the gas produced there for conveyance.

Jasmine Field

Oil field in the UK sector of the North Sea, undergoing development in the expectation that production will begin in 2012. The water depth is 81 m. The rig used in exploratory drilling was Ensco 102, a **jack-up** device. Two sidetrack wells were drilled and 167 m of **pay** was discovered indicating quite a major field. There will be a wellhead platform, and oil from Jasmine will be taken to the platform at the **Judy Field** 9 km away. This is at the same water depth as the Jasmine Field, and is operated by Phillips Petroleum. (*See* **Janice Field**.)

Jasper Explorer

Drill ship, having undergone major upgrading in 2009. It can drill in water depths up to 1500 m and to well depths of 7000 m. It has **dynamic positioning**, the fuel requirement of which is 45 cubic metres of fuel oil per day. This is examined in the box below.

$$\text{Rate of heat release} = [45\ \text{m}^3 \times 900\ \text{kg m}^{-3} \times 43 \times 10^6\ \text{J kg}^{-1}/(24 \times 3600)]\ \text{W}$$

=20 000 kW approx.

For a conversion of say 35% of heat to work, the rating of the DP becomes:

7000 kW = 9400 h.p.

Jasper Explorer is scheduled in the near future for duty offshore the Republic of Guinea.

Jatibarang Field

In Java, Indonesia. Over part of its 500 square kilometre area the field comprises coal ('organic rock'), and therefore **CBM**, in addition to formation containing oil. Production of oil there began in 1973, and it has been judged that several new horizontal wells will be needed in order to improve the output from the field. The Jatibarang Field features in Indonesia's CBM programme but not as strongly as, for example, the **Barito Basin**. (*See* **Huabei Field**.)

Jeanne d'Arc Basin

Offshore Newfoundland, the scene of oil production since 1997. The water depth is 80 m. Current production is 126 000 barrels per day and the operator is ExxonMobil. A major factor in production platform construction was the ability to withstand iceberg impact. The platform now in service at Jeanne d'Arc is a **GBS** and has been named the Hibernia platform. It contains a storage tank which receives oil from the wells and can hold up to 1.3 million barrels. Hibernia is often considered to be the largest oil platform in the world. There is a caisson, the upper surface of which is just above the waterline, and this is enclosed by an 'ice wall' having sixteen 'teeth'. Quite simply these will penetrate any iceberg which strikes the platform and convert its kinetic energy into thermal where the 'teeth' and iceberg contact. (*See also* **Shtokman gas condensate**; **Terra Nova Field**; **Yolla-A platform**.)

Jebel Ali Refinery

Condensate refinery in Dubai operated by **ENOC**. It began operations in 1999. Its production in 2010 was 108 000 barrels per day. The nominal capacity is 120 000 barrels per day. The refinery receives **stabilised condensate** from Qatar, Iran and Australia. There is a pipeline which takes jet fuel produced at the refinery to the airport in Dubai. Naphtha, lighter than jet fuel, is exported as chemical feedstock and the remainder of the distillate finds application within the UAE. Vessels delivering stabilised condensate for processing at Jebel Ali can be up to 120 000 tonnes deadweight, and this converts to approximately:

$(120000 \times 10^3\ \text{kg}/700\ \text{kg m}^{-3})/0.159\ \text{m}^3\ \text{bbl}^{-1} = 1$ million barrels

and a vessel capable of holding that would most probably be classified as a 'supertanker'. (*See* **J.F.J. de Nul**; **Lamerd gas condensate refinery**.)

J.F.J. de Nul

The largest **cutter suction dredger** currently in operation, having entered service in 2003. Its scenes of activity have included **Jebel Ali Refinery**, where it operated at a **BCD** of – 17 m. It has also seen service at Sakhalin. Its range of water depths is 6 m to 35 m. Its specifications have been studied by the author and some *very rough* calculations follow.

The cutter drive power of J.F.J. de Nul is 6 MW, which we can take to include the power requirements of suction. Now for the *particular case* where the seabed is of loose sedimentary rock, no breakage of which is required this power to be expended, to a fair approximation, only in vertical movement. Working at its maximum water depth of 35 m, the rate ***m*** at which it will lift sea-floor debris to level **BCD** zero is calculable from:

$$\boldsymbol{m}\ \text{kg s}^{-1} \times 9.81\ \text{m s}^{-2} \times 35\ \text{m} = 6 \times 106\ \text{J s}^{-1}$$

$$\downarrow$$

$$\boldsymbol{m} = 17\ \text{tonne per second.}$$

A performance of 'half a million cubic metres of debris per week is the basis of the following approximate calculation. Using an arbitrary but reasonable figure of 2000 kg m^{-3} for the bulk density of the sedimentary rock charged with seawater, the figure above for J.F.J. de Nul becomes:

$$\{(17 \times 10^3\ \text{kg s}^{-1}/2000\ \text{kg m}^{-3}) \times 24\ \text{h d}^{-1} \times 3600\ \text{s h}^{-1} \times 7\ \text{day-week}^{-1}\}\ \text{m}^3\text{week}^{-1}$$

$$= 5\ \text{million cubic metres per week.}$$

The two values differ by just one order of magnitude and this reflects two factors: the scale of the J.F.J. de Nul and the assumption of loose sedimentary rock. It can reasonably be asserted that there is consistency in the results[34]. Of course, in the event that J.D.F. de Nul was working at a very loose seabed, full power need not have been engaged. The suction pipe on J.F.J. de Nul is 1000 mm in diameter. The dredger has travelled widely between jobs, and has done so at about 2.5 knots under its own propulsion at 5000 h.p.

Jidong Field

Onshore oil field in northern China operated by Petrochina having over a billion barrels of recoverable reserves. The field has been the scene of **Frac-Pack completions**. (*See* **Centerfire™**.)

Jotun

A.k.a. Jotun A, a new-build **FPSO** having entered service in 1999 and in long-term operation at the field of the same name – the Jotun Field – in the North Sea at a water depth of 126 m. It has **turret mooring** and can therefore 'weather vane'. Its peak production over its years of activity was 145 000 barrels per day and it can hold 0.58 million barrels of crude oil offloadable at 40 000 barrels per hour. There is also provision for **water injection**. The formation at Jotun Field has **pay** ranging from 18 m to 46 m at depths of about 2000 m below the seabed.

Jubilee Field

Offshore Ghana, water depth 1250 m. The operator is Tullow oil. Oil production is at an **FPSO** named Kwame Nkrumah MV21, a rebuild having previously been a tanker. It has **turret mooring** and can store up to 1.6 million barrels of oil. The first exploration well revealed the presence of a **gas cap**. Weatherford have entered into a contract at the field whereby they will carry out operations including pigging.

Judy Field

Condensate field in the UK sector of the North Sea, producing since 1995. The water depth is 75 m and production is at a platform attached to the sea-floor by 14 piles each of length 115 m driven to a depth of 92.5 m. The formation is chiefly sandstone with permeabilities up to 1100 millidarcys and porosities of typically 27%.

Juliet Field

Gas field in the southern part of the UK sector of the North Sea, currently being developed. The formation is of good permeability and it is planned that there will be two horizontal production wells tied back to a platform operated by BP at the Amethyst Field east of Juliet. From there the gas will be transferred to **Dimlington**.

Jumper

Subsea connection piece for oil transfer. (*See* **Casablanca Field**.)

Kapap Natuna

FPSO in service offshore Indonesia operated by Star Energy. Its mooring ties are attached to the vessel by a yoke. It receives oil from ten production wells at a rate of 25 000 barrels per day.

Karachaganak Field

In north-west Kazakhstan, producing oil, gas and condensate. Companies involved have included British Gas, Eni, Chevron and Lukoil. Production began in 1984 and peak production in the 1990s was 4.3 billion cubic metres of gas per year and 100 000 barrels of liquid per day. Expansion occurred in the early 21st Century and this included a 650 km pipeline to **Atyrau** providing access to the **Caspian pipeline consortium** and a route to export by pipeline to Russia. Some is taken to China via the **Central Asia-China gas pipeline**. Advances at the field since then include workover of wells previously in abandoned status and the drilling of new exploration and production wells. (*See* **Sorochinsko-Nikolskoe (S-N) Field**.)

Karoo Basin

In South Africa, covering an area of 88 000 square kilometres mostly in the Orange Free State. It is believed to contain major amounts of **shale natural gas**. South Africa's own oil company Sasol are working with Statoil and Chesapeake Energy in evaluation. Shell obtained an exploration licence in 2009[35]. A few exploration wells had been drilled at the Karoo Basin in the 1960s and the 1970s, and each gave evidence of **pay**. It would have been exceedingly difficult to organise an international team to develop a resource in the South Africa of that time.

Kårstø processing plant

Near Stavanger, Norway, receiving gas from fields including Åsgard. **Statpipe** also delivers gas to the Kårstø facility. Condensibles are removed and refining to ethane, propane, n-butane, i-butane and naphtha takes place.

The remainder is **stabilised condensate** and passed along as such. The dry gas goes via Europipe to Germany. (*See also* **Åsgard pipeline**; **Nam Con Son Field**; **Njord Field**; **Skarv**.)

Kashagan Field

In very shallow (<5 m) of water offshore Kazakhstan, commencement of production expected in the near future. Oil companies involved in the development have included Shell, ExxonMobil, ConocoPhillips and Eni. Oil field services companies contributing have included Halliburton. Initial production will be from three wells. The formation at the field is carbonate. The reservoir depth is 4500 m and the **pay** of oil is over 1000 m. Drilling of the production wells was from the **Sunkar floating production vessel**[36]. There are offshore structures ('hubs') for processes including stabilisation of crude oil. Oil and any associated gas which is not reinjected or used at the facility will be taken by pipeline to the Bolashak terminal onshsore. Initial production at the field will be 75 000 barrels per day rising to 1.5 million barrels per day over a ten-year period of concurrent production and expansion. (*See also* **Tengiz**; **Tolkyn Field**; **Zhanazhol Field**.)

Kaskida Field

In the Gulf of Mexico and operated by BP. The water depth is 1800 m. The exploration well is 9900 m deep, representing a drilling record in the Gulf. There are as yet no production wells. The **pay** revealed by the exploration well is also high, about 240 m. Drilling of the exploration well was by the **semi-submersible rig** West Sirius. This can operate in water depths of up to 3050 m and can drill to depths of 11 500 m below the sea-floor. Its **derrick** can hold a load of 1100 tonnes and its **drawworks** are rated at 4500 h.p. (*See* **Telemark Hub**.)

Kazakhstan-China oil pipeline

This facility, the name of which is self-explanatory, was built in a number of segments. The first, completed in 2002, runs within Kazakhstan from **Atyrau** to Aktobe. Aktobe is the scene of oil fields including Kenkiyak, which is operated by China National Petroleum Company. Atyrau is on the Caspian coast and is well set up with oil production 'paraphernalia', receiving from on- and offshore fields. When this part of the pipeline was initially used its purpose was solely domestic, taking oil east from Kenkiyak to the coast for processing. Now that it is part of the Kazakhstan-China oil pipeline, its contents flow in the opposite direction, taking oil west to east, that is, in the direction of the border between Kazkhstan and China. This segment is of length 449 km and diameter 24 inches. Oil departing this first segment originates from offshore fields having been brought in at Atyrau, from Kenkiyak and also from the **Zhanazhol Field**. The next phase of construction was the Kenkiyak-Kumol segment of 793 km length and 32 inch diameter. Oil from the field at Kumol can be admitted to the pipeline here. The final part is the Atasu-Alashankou segment of length 963 km and diameter 32 inches. Its termination – Alashankou – is at the border between China and Kazakhstan. According to the above figures the total length is:

$$[449\ \textit{Atyrau-Aktobe} + 793\ \textit{Kenkiyak-Kumol} + 963\ \textit{Atasu-Alashankou}]\ \text{km}$$

$$= 2205\ \text{km}$$

In 2010 the pipeline exported approximately 70 million barrels of oil from Kazakhastan to China. At one of its junctions the pipeline changes from 24 inches to 32 inches diameter, but the mass flow rate must by the principle of continuity be the same in pipe of either diameter. The speed in the 24 inch (0.61 m) segment can therefore be estimated in the following way.

$$70 \times 10^6\ \text{barrels year}^{-1} = 70 \times 10^6\ \text{bbl} \times 0.159\ \text{m}^3\ \text{bbl}^{-1}/(365 \times 24)\ \text{hours year}^{-1}\ \text{m}^3\ \text{h}^{-1}$$

$$= 1270\ \text{m}^3\ \text{h}^{-1}$$

$$\text{Flow speed} = \text{volumetric flow rate/pipe cross section} =$$

$$[1270\ \text{m}^3\ \text{h}^{-1}/(\pi \times 0.305^2)\ \text{m}^2] \times 10^{-3}\ \text{km h}^{-1} = 4.3\ \text{km h}^{-1}\ (2.7\ \text{m.p.h.})$$

The value obtained above is at the low end of the range of speeds of oil in pipelines – 3 m.p.h. to 8 m.p.h. is the 'rule of thumb' range often quoted – and it will be lower still in the 32 inch pipine. This will have to do with the **pumping stations**, their number and whether some or all of them are in use at any one time. (*See also* **Central Asia-China gas pipeline**; **Myanmar-China pipeline project**; **Karachaganak Field**.)

Keith Field

Gas and condensate field in the UK sector of the North Sea, tied back to infrastructure at the nearby **Bruce Field**. It is operated by BHP Billiton. The liquids yield is 2150 barrels per day. The component of the liquids remaining after condensate stabilisation becomes LPG which, of course, is standard procedure.

Kemerkol project

Exploration in Western Kazakhstan, operator Victoria Oil and Gas plc (HQ in London) and close to **Atyrau**. Oil was first produced there in 2006.

Legal challenges concerning the ownership of the field, referred to the Supreme Court of the Republic of Kazakhstan, have delayed development. Prior to the legal difficulty target production was 4000 barrels per day. (*See* **Logbaba**.)

Kenai LNG plant

In Alaska, the only LNG production facility in the US from which export occurs: all of those in the more southerly 48 states produce for the domestic market only. The joint owners are Conoco Phillips and Marathon but Marathon is the operator. The gas liquefied at Kenai is from **Cook Inlet**. The Kenai plant has been in operation since 1969 and its major customers have been Tokyo Electric and Tokyo Gas. The year 2009 was a 'high' for those involved with the plant, marking the 40th anniversary. Alas in February 2011 the imminent closure of the plant was announced[37]. The assets of the plant had included two LNG tankers, reduced to one in 2010. Each had been built in Japan and had been in use for over 15 years. (*See* **Gladstone LNG project**.)

Kharg Island, Iran

Site of the largest oil terminal in the Persian Gulf. There are also extensive storage facilities (>20 million barrels), a refinery and petrochemical plants. Kharg Island is one of the major hydrocarbon centres of the world. In this entry a new activity there, called the *Kharg Island Gas Gathering and Natural Gas Liquids Recovery Project*, has been selected by the author for detailed coverage. For this about 30 million cubic metres of gas per day will be taken to Kharg from onshore fields and an equivalent amount from offshore. The incentive for the project is to reduce the degree of flaring of associated gas which currently takes place at Iranian oil fields. Compression units, two of them expressly commissioned for the project, will take the gas to Kharg Island and there will be desulphurisation as necessary. Products of the 'recovery' will be gaseous methane, propane, butane, pentane. What remains has the nature of **stabilised condensate**. The lighter products will be directed towards the petrochemical industry at Kharg. The miscellany of useful products are an entire bonus: the gas they came from would, as previously noted, simply have been flared.

Kharyaga Field

In the part of Russia occupied by the Nenet (Uralic) race, an onshore field in a permafrost setting. There had prospecting much earlier, but production did not begin until 1999 since which time expansion has been considerable. There are now fourteen wells, and production is 30 000 barrels per day. This is all taken by pipeline to a location on the Baltic coast close to

St. Petersburg. It has been necessary to flare associated gas, but the next phase of development involves introduction of gas processing in order that the gas be recovered for marketing.

Khazar

Semi-submersible rig under construction in an Iranian shipyard and for use exclusively in the Iranian sector of the Caspian Sea. Much of the Caspian Sea is shallow - some of it less than 10 m - and the deepest parts are in the Iranian sector. Even here depths are modest, and the Khazar will be able to drill at depths of up to 1070 m. It will have conventional mooring. There are two other semi-submersible rigs in the Caspian Sea at present, each with a depth limit slightly less than 500 m. Once the production wells have been drilled it is likely that a semi-submersible rig will be used as a production rather than as a drilling base, as at the much deeper **Independence Hub**. (*See* **Khvalynskoye Field**.)

Khvalynskoye Field

In the Caspian Sea, discovered in 2002 and under development with a view to gas and condensate production by 2016. Typically of the Caspian the water depth is shallow, actually 25 m to 30 m, and this makes the **jack-up** rig suitable for use in drilling. Target initial production is 25 million cubic metres of gas per day plus condensate. (*See* **Harding Field**.)

Kick

Uncontrolled entry of fluid into a well bore. If the fluid is hydrocarbon, kick can be the precursor to blowout, necessitating activation of a **BOP**. Kick involving drilling mud can be addressed by adjusting the density of the drilling mud.

Kikeh

Oil field offshore Malaysia currently being developed. The water depth at the site of proposed production is 1350 m. Well drilling from a spar unit is proposed. Oil from wells so drilled will go to an **FPSO** having an **external turret** and there will be a line for **water injection**. The field is believed to contain 700 million barrels of crude oil.

Kinsale Head Field

Gas field offshore Ireland, discovered in 1971 and producing since 1978. The previous year two **fixed platforms** - Alpha and Bravo - were installed at the field. The water depth is 100 m and subsea drilling is to a depth of 1000 m. The operator is Marathon. The most productive year over the field's history was 1995 when the field yielded 2.7 billion cubic metres.

Kipper Field

In the Bass Strait offshore Australia, potentially productive of natural gas and of condensate, discovered in 1986. Commencement of production is imminent at the time of going to press. There will initially be two production wells, tied back to existing infrastructure at Longford.

Kiskadee Field

Offshore Trinidad and Tobago, productive of gas and condensate. Discovery was by a group led by Texaco in 1977. Production began in 1993 at an installation also set up for drilling. The installation stands in 69 metres of water and has a capacity of 4 million cubic metres of gas per day and 10 000 barrels of condensate.

Kitimat LNG terminal

Under construction in British Columbia, expected to be operating from 2013. It has been explained how many proposed LNG terminals in the western USA have come to nothing, and Kitimat and the **Costa Azul LNG terminal** will be the only such facilities on the west coast of North America and neither is in the US! The amount of LNG regasified will be 5 million tonnes per year about a quarter of which will have come from Australia. The choice of location of the terminal is such that it will be suitable for import of LNG from Sakhalin sooner or later. (*See* **St John, New Brunswick**.)

Kittiwake

Oil field in the North Sea, and until recently the scene of an offloading buoy (a.k.a. an **FSO**) of the same name. The buoy has been decommissioned, a process which began with its towing to the *Vats Decommissioning Yard* in Norway. There it was cut into four pieces unequal in weight. Three of them each had a weight of about 800 tonnes, and were lifted out of the sea by a particular crane; another crane had to be deployed for the remaining piece, which weighed almost 2500 tonnes. The moieties having been so brought into the decommissioning yard were broken up further and the pieces examined for possible recycling or disposal, compliance with legislation being the paramount factor in either case. Flexible flowlines which had previously conveyed oil on and off the Kittiwake buoy were retained for re-use.

Kizomba

Oil field development offshore Angola and currently divided into Kizomba A, Kizomba B and Kizomba C. At Kizomba A the water depth is 1189 m and there is both a **tension leg platform** and an **FPSO** and they are linked to each other by fluid transfer lines. Kizomba B (water depth 1020 m) also uses these two types of production facility and there are two FPSO vessels at Kizomba C, where the water depth is 730 m. Oil is transferred from the production facilities to an oil offloading buoy (a synonym for **FSO**) by means of steel **OOLs**. (*See* **Knock Adoon**, **Kittiwake**.)

Knock Adoon

FPSO operating in Nigerian waters under an eight-year contract between the owner Olsen and Addax Petroleum. It is a conversion from a tanker built by Mitsubishi and has **spread mooring**. Oil is transferred from it to an offloading buoy. The storage capacity is 1.7 million barrels of crude oil. (*See* **Sangaw North**.)

Kochi Port

In southern India and the site of a refinery. The port is of sea depth only 12 m, precluding the entry of oil tankers, so oil is transferred from a tanker to a loading buoy with SPM. A pipeline was installed from the buoy to the mainland and this necessitated dredging. (*See also* **Cauvery-Palar basin**; **HAM 318**.)

Kollsnes

Plant in Norway for **condensate stripping**, taking gas from **Troll**, **Kvitebjørn** and **Visund**. Condensate produced is taken to the Mongstad refinery. This is actually a 'conventional refinery' set up to fractionate crude oil, but can receive condensate. (This is not an uncommon arrangement: see **Bagaja gas condensate field**). Kollsnes can receive up to 143 million cubic metres of natural gas per day for processing which will release 69 000 barrels per day of condensate for transfer to Mongstad. The gas after stripping is admitted to **Statpipe**, **Zeepipe I** and Zeepipe IIA. (*See also* **Gjøa Field**; **Pioneer Knutsen**; **Zeepipe I**.)

Kome Kribi 1

FSO with **tower mooring** in service offshore West Africa. In very shallow water (34 m), it receives oil from the **Doba Fields** in Chad and retains it for transfer to tankers. It has been in service since 2002. It was not purpose built but was converted (in Singapore) from an Ultra Large Crude Carrier (ULCC) and can hold over 2 million barrels of crude oil.

Kovytka Field

In eastern Siberia, recently acquired by Gazprom. It is intended that gas from the field will be exported to China. Previously, lack of a local market for gas from the field had made its development by TNK-BP non-viable. (*See* **Urengoy Field**.)

Kravtsovskoye (D-6) Field

Oil field in the Baltic Sea in 35 m of water, operator Lukoil. There is a **fixed platform** at the field with multiple wells at depths of about 2150 m. Production began in 2004 and by 2007 oil yields of the order of 0.25 million barrels per day were being realised.

Kudu gas field

Offshore Namibia, water depth 171 m. Expected commencement of production 2014. The operator is Tullow Oil. The **pay** is 15 m and by early 2010 there were seven wells at the field. It is proposed not to collect the gas for sale but to burn it to make electricity some of which will be sold to South Africa.

Kumul Marine Terminal

In the Gulf of Papua, receiving crude oil by pipeline from Kutubu, Moran and Gobe fields. From the terminal oil is offloaded onto tankers berthed by SPM at a rate of 200 000 barrels per day up to a maximum tanker capacity, set by the sea depth, of 650 000 barrels. (*See* **Kutubu Oil Project**.)

Kutubu Oil Project

The first commercial oil field development in Papua New Guinea, in the southern region of that country. Oil, in a sandstone formation, was first discovered there in 1986 and production began in 1992. There are 46 wells at the field, some of them for production and some for gas reinjection. The Kutubu Field (which takes in the 'Agogo') has recently performed at well below its initial production rate, partly because of a high gas-to-oil ratio. Production for 2010 was 3.66 million barrels, down an order of magnitude on the figure from the early 1990s. New wells are being drilled, and development is of Kutubu jointly with the nearby Moran and Gobe fields. Oil from Kutubu, Moran and Gobe is sent by pipeline to the **Kumul Marine Terminal**. Blended oil from the three fields is marketed as 'Kutubu light'. (*See* **PNG LNG project**.)

Kvitebjørn

Tension leg platform in the Norwegian sector of the North Sea standing in 190 m of water having received gas and condensate since 2004. It has drilling as well as production capability and was used to drill the appraisal well at the nearby **Valemon Field**. Obviously a fixed drilling facility will need to be set up for **directional drilling** necessitating, for example, a **downhole motor** to orientate the drill bit. (*See* **Kollsnes**.)

Kyoto tanker

(*See* **Pioneer Knutsen**.)

Laggan and Tormore fields

North west of Shetland, a condensate field in 600 m of water. The gas/condensate is at depths of about 3500 m below the sea-floor. Distance from existing infrastructure will be a major influence in the planning of production. Nevertheless, it is expected by Total that gas production will begin in 2014.

Lake Maracaibo

In Venezuela, in the early 1990s the scene of recovery of **attic oil**. This was by the drilling of a bore in the 'attic sections' of the reservoir. The drilling trajectory was planned having regard to the presence of a **fault**. This operation at the first attempt was unsuccessful because of what is described in an SPE conference paper on Lake Maracaibo[38] as 'the complexity of the geologic structure'. Advances in **LWD** might have eliminated this difficulty had the work been undertaken a decade or so later. Even so, a second drilling operation *was* successful and the well when finally constructed was more productive than any of the existing ones at the reservoir. (*See* **Bolivar Coastal Field**.)

Lamerd gas condensate refinery

In Iran, commissioned in 2005. It receives condensate from the South Pars Field in Qatar. The hoped for eventual performance is refining of 120 000 barrels per day of condensate. Gasoline and jet fuel are amongst the products. The former will go to the regional airport in Lamerd from where there are regular flights to Tehran. (*See also* **Jebel Ali Refinery**; **Tolkyn Field**.)

Laminaria Field

In the Timor Sea, operated by the Australian concern Woodside Petroleum and the Canadian concern Talisman. Production is at **Northern Endeavour** which receives from eight production wells. The water depth at Laminaria is 340 m, and the location of the field such that whether it is in Australian or Indonesian waters has become a matter of debate. (*See* **Sunrise Field**.)

Laos, oil exploration in

Oil production in Laos at the present time is nil. The UK company Salamander Energy, which specialises in Far East activity, is drilling for oil in Laos with drills previously in use in Thailand. The Saravane Province of Laos, in the southern part of the country, is the scene of drilling of exploration wells between 3000 m and 4000 m deep. It is reported that the drilling of one such well with the drill from Thailand will take 2 to 4 months, which, if drilling is for eight hours in the day for five days in the week, signifies an **ROP** of:

$$10\,000 \text{ feet}/(0.3333 \times 12 \times 5 \times 24) \text{ feet per hour}$$

$$= 21 \text{ feet per hour } (6.4 \text{ m per hour})$$

which is a fairly standard value. 'Shows' of oil were reported in Laos in the 1930s, and much more recent endeavours have included the **spudding in** of a well ('Paske-1') in 1996 by Hunt Oil. This was later revealed to be a dry hole. (*See also* **Bualuang Field**; **Sinphuhorm gas field**; **Toroa well**.)

Las Cienegas Field

Oil field close to the centre of Los Angeles. Many of the wells there had to be shut down in the late 1990s because they were in such poor shape. There had been significant clay deposition at these wells, also carbonate decomposition. In the early 2000s measures were taken to 'acid stimulate' five of the shut down wells to restore their viability. Use of **mud acid** resulted in recovery of production only for a matter of days before the wells were again very restricted. A stimulating agent containing both hydrochloric acid and, less conventionally, phosphoric acid was therefore developed and used at the five wells at Las Cienegas. By March 2005 all of the five wells were producing. The undertaking was made more realistic by the large increase in the price of oil not only over the period of their shutdown but in the couple of years following their being brought back into use.

LB 200

Pipe laying vessel operated by Stolt Offshore (now Acergy) using **S-Lay**. It is sometimes compared with **Solitaire** and such comparison is sound in terms of scale. The important difference is that LB 200 is moored by anchors and cables whereas Solitaire has **dynamic positioning**. The LB 200 has in fact been in service much longer than the Solitaire. The obvious reason why those in the industry have tended in recent years to compare LB 200 with Solitaire is that both vessels played a major part in laying the pipeline from Nyhamna in Norway to Langeled on the east coast of England – a 1200 km pipeline – in 2005-2006 (the 'Langeled pipeline'). In this undertaking

LB 200 started its pipe laying work at the Sleipner platform in the Norwegian sector of the North Sea, working in the direction of Langeled. Solitaire began its duty at Langeled and then moved to Nyhmana to commence pipe laying work there. One effect of the Langeled pipeline when it came into service in late 2006 was to make the UK for a period a net importer of natural gas, having previously been a net exporter. (*See also* **Ormen Lange Field**; **Scarab and Saffron Fields**; **Statpipe**.)

L-Gel

Gel pig from Aubin Subsea. Its unusual feature is its elasticity. If along its route there is partial blockage reducing the effective diameter of the pipe, the **pig** will, on exiting that part of the pipe, return to full pipe diameter. The manufacturer's web pages describe this effect as a 'gel pig with a memory'. It has been used in the North Sea in commissioning and decommissioning of pipelines. MEG Gel, from the same manufacturer, is often placed in front of L-Gel and carried along by it.

Leighton Eclipse

Newly built pipe laying barge having made its debut at the Mumbai High Field where it laid 2820 m, comprising 235 sections of 12 m length, in one day. The pipe so laid is 16 inches in diameter. This is part of a pipeline which will eventually extend to 80 km so at the rate of the initial lay this will take:

$$80000 \text{ m}/2820 \text{ m d}^{-1} = 28 \text{ d}$$

The pipeline being laid by Leighton Eclipse is a replacement of an existing pipeline.

Leiv Eiriksson

Semi-submersible rig, built in China and entering service in 2001. It has been in use off Angola and can operate in water depths of up to 2300 m, making the claim by its operator Ocean Rig that it is for 'ultra deep water' use quite sustainable. In 2011 Leiv Eiriksson was departing Istanbul for Greenland when Greenpeace 'activists' boarded it and placed on its **derrick** a banner reading 'Stop Arctic Destruction'. The rig continued on its way. The rig is to operate close to Greenland on behalf of Cairn Energy, at the invitation of the government of Greenland. (See **Disko Island**.)

Leonardite

Substance intermediate between peat and lignite having a higher content of humic acids than a lignite would. Occurring in places including New Mexico USA, it can be used in the production of drilling muds enabling the viscosity of a particular mud to be reduced. Leonardite might be processed for such use, as with **HUMIN P 775**. Leonardite and other humic substances can be used in **filtration** control. Where leonardite is not available, lignites – brown coals – can be used to make humic products for **drilling fluids** and other applications such as soil treatment. As already noted they are lower in humic acids than leonardite but not so much so as to preclude such use. This has sometimes led to synonymous use of the terms 'leonardite' and 'lignite' and as long as this is understood as only applying to use of the respective substances in drilling mud manufacture this is acceptable. The two terms are not of course synonymous in other senses. That lignites are hydrophilic is clear from the form in which they occur naturally. Some in bed-moist state are 65% moisture and the term 'giant sponge' is sometimes applied to a deposit of lignite. This property is part of their function in some drilling muds. Leonardite can be treated so as to make it organophilic. In this form it is used in oil based drilling muds as a filtration control agent, for example **DURATONE® E**.

Leviathan Field

Gas 'prospect' in the Mediterranean. It is in Israeli waters, although Lebanon is vigorously asserting a claim at least to part of it. The water depth is 1634 m. The **semi-submersible rig** Sedco Express commenced drilling an exploration well there in October 2010. Results are awaited and might be disappointing, but of the order of 450 billion cubic metres of natural gas is hoped for. This is thermally equivalent to 2.5 to 3 billion barrels of crude oil, the amount at the **Tiber Field**[39]. Other Israeli oil fields including the **Tamar Field** will add not insignificantly to this. If these finds of gas enable Israel to become self-sufficient in energy that will affect the very delicate relationship which the country has with its oil-producing neighbours.

Liquefied natural gas (LNG), manufacture of

LNG is a staple fuel and at any one time there are many LNG tankers on the oceans. The world's largest importer of LNG is Japan and the world's largest exporter is Malaysia. The USA exports LNG only from Alaska and that is about to cease. The method of manufacture of LNG is as follows. Gas (which, if it was associated gas, will previously have been separated from crude oil) is treated to remove impurities such as carbon dioxide and hydrogen sulphide using such substances as **MEA**. It is then cooled by refrigeration to remove condensate and refrigerated further to −160 °C which is the normal boiling point of methane. At this stage it becomes a cryogenic liquid in equilibrium with its own vapour at 1 bar, and that is the product known as LNG. There are a number of technologies for making LNG and they differ in the details of refrigeration. Mixed refrigerants are common and some contain liquid nitrogen amongst other ingredients. Obviously heat exchangers feature in each of the refrigeration processes and heat exchanger configuration and performance are a major feature of any one of the processes. For example, the spiral wound heat exchanger (SWHE) made by Linde for LNG manufacture is a proprietary product. It is used in the liquefaction step only in conjunction with a simple parallel flow heat exchanger for the first refrigeration step. The path from raw gas to LNG product is an *LNG train*, and there will often be multiple 'trains' at an LNG liquefaction plant. The Karratha liquefaction facility in NW Australia, which receives unprocessed gas from Cossack Pioneer, has seven LNG trains. Different trains at the same place will not necessarily have the same operator or use the same refrigeration methods. This is how the LNG facilities of today's world work. An oil refinery might also have one or more LNG trains even though LNG is not its primary business. LNG has in fact been in existence for over 60 years, and the earliest plants used the Joule-Thomson effect to cool the gas by passing it through an orifice, thereby converting thermal energy to kinetic sufficiently to bring the methane below its critical temperature. (*See also* **Kenai LNG plant**; **Peru LNG Project**; **Sakhalin**; **LNG production at**; **Snøhvit LNG facility**; **RMU**; **Yemen, LNG production in**.)

Liverpool Bay offshore storage installation

Tanker permanently moored in the Irish Sea having a capacity of 0.87 million barrels of crude oil, operated by BHP Billiton. It is double hulled and positioned so as to be outside shipping lanes. It receives oil by pipeline and offloads it on to **shuttle tankers**. The tanker is moored to a buoy, and tension monitoring of the cables connecting the two is carried out on a continuous basis. This is necessitated by the fact that the tanker moves through a full circle, with the buoy at its centre, twice during a typical 24 hour period because of tidal movements.

LL652 Field

Gas field in **Lake Maracaibo**, operated by Petroindependiente of which 25.2% is owned by Chevron. Daily production is 1.4 million cubic metres of gas and 5000 barrels of condensate. There is a production platform at the field, and also platforms for **water injection** and gas compression. These were made in the US and taken by barge to the Port of Maracaibo. Attachment to the lake floor is by piles 300 feet in length. (*See* **Zuata Field**.)

Lloyd Noble

Jack-up rig, capable of operating in water depths up to 76 m and of drilling wells at depths up to 6000 m. It was built in 1983 and had a rebuild in 1990. Its **POB** is 96. Its **drawworks** are rated at 2000 h.p. (*See* **Akepo Field**.)

LNG

(*See* **liquefied natural gas**)

Logan prospect

In the Gulf of Mexico, the scene of proposed drilling by Statoil. The water depth is 2381 m, significantly deeper than the Macondo prospect where there was a major spill in 2010. (*See* **Discoverer Americas**.)

Logbaba

Gas and condensate development in Cameroon, in which Victoria Oil and gas plc are a major participant. It is in the Douala Basin. **Pay** of hydrocarbon is at depths of 2135 m to 2590 m in a sandstone formation, and further pay occurs at an extra drilling distance of about 1000 m there being a *barren interval* in between. Exploratory drilling was in fact to a **TVD** of 4172 m. Once fully operational the field will supply gas for electricity generation for urban Douala. (*See* **Sapele well**.)

Lomond Field

Gas field in the North Sea. A **fixed platform** there receives gas from six production wells. The platform also receives inventory from the Erskine Field and will in the near future receive inventory from the **Columbus Field**. (*See also* **Cottonwood Field**; **Durango well**.)

Long-distance subsea tieback

(*See* **Cottonwood Field**.)

Longhorn Field

Gas field in the Gulf of Mexico, production having commenced in mid-2009. The water depth is 730 m and production is tied back to the Corral platform. The major owner is Eni. Drilling vessels deployed at the field were **Amirante** and Ocean America, one of a family of **semi-submersible rigs** owned and operated by Diamond Offshore. (*See also* **Appaloosa Field**; **Firenze**.)

Lorelay

Pipe laying vessel from the AllSeas fleet, the first such vessel to have **dynamic positioning**. It has laid pipes at water depths up to 1645 m. Lorelay has seen recent service in the Mississippi Canyon where the **S-Lay** approach was used in spite of the very considerable water depth. This was made easier by the ability to 'position dynamically'.

Lost circulation

Drilling mud having fulfilled its function at the drilling bit is expected to enter an annulus. Lost circulation is the state of affairs whereby the fluid enters spaces and voids in the rock structure (formation) instead. A bridging agent can be used to close off the channels into which the **drilling fluid** has been diverted and redirect it to the annulus. Common bridging agents are calcium carbonate (inorganic) and ground cellulose (organic), e.g. **All Seal**. **Benzoic acid** is also so used, in flake form. Sometimes when the loss is through a network of fine channels a fracture is intentionally made in the formation to allow admittance of an agent which will enter the channels and then consolidate (e.g. **STEELSEAL®**), so closing them. This will be followed by use of a conventional bridging agent. Excessive entry of drilling mud into the annulus through a pressure surge is an example of **kick**.

LTMO

Low-toxicity mineral oil, a term for particular petroleum derived base oils for **drilling fluid** use.

Lufeng 13-1

Offshore China, scene of operation of a **disconnectable turret**. This was made by adaptation (in Singapore) of an existing vessel and can hold 0.88 million barrels of oil. The water depth is 75 m.

Lula Field

Formerly named the **Tupi Field**, oil and gas field offshore Brazil discovered in 2006. The water depth is 2150 m and the oil is 2000 m subsea with salt rock intervening. Conditions are therefore challenging, but amounts of oil believed to be present are sufficient for production to be authorised, as is indeed the case. One **FPSO** called Cidade Angra dos Reis is already moored at Lula having commenced production in October 2010. This has **spread mooring** and receives from 5 wells. Many more production facilities will be put in place at Lula and contracts for new-build FPSOs have been awarded. Full implementation of the projected production activity will not be until 2017 and flexible pipeline manufactured by Technip will feature in the developed field. (*See also* **Acergy Condor**; **Carbon dioxide storage in oil fields, comments on**.)

Lumut

In Brunei, the site of an LNG plant. There are five LNG trains with a sixth, larger than the existing ones, planned. The current capacity is acceptance of 5.3 million cubic metres of gas per day from from offshore Brunei for liquefaction. The NGLs (natural gas liquids) present are separated into pure compounds (ethane, propane) and some of this is put to refrigerant use at the plant. The separator is therefore referred to as an **RMU**. The LNG from Lumut is exported to Japan and to Korea. (*See also* **Ampa Field**; **Champion Field**; **Rasau Field**.)

Luno Field

Oil field in the Norwegian sector of the North Sea under development. It is further out to sea than, for example, the **Glitne Field** and the **Sigyn Field**. Close to Luno is the Daphne Field, and joint development is planned. Exploration drilling at Luno in late 2007 revealed oil, and appraisal wells followed. A further exploration well on the south side of Luno in 2009 also gave very positive results. At Daphne **spudding in** of an exploration well took place in late 2010 from the **semi-submersible rig** Transocean Winner. The target subsea depth is 2500 m.

Lunskoye-A

Platform offshore Sakhalin Island. The **perforating gun** used in the **pay** zone was operated by **TCP** and contained PowerJet explosive charges, which is a Schlumberger product. These provided 12 shots per foot (39 shots per metre) at casing depths up to 2200 m.

LWD

Logging whilst drilling. (*See also* **Meji Field**; **MWD**.)

Maari Wellhead Platform

Offshore New Zealand, one of two sites of Arup's DrillACE structures, whereby a gravity base structure is tied to an **FPSO**. Because of the absence in a gravity base structure of any sort of attachment device such as piles, the term 'self-installing' has been applied. At Maari the water depth is 100 m and production is 35 000 barrels per day of oil. (*See* **Yolla-A platform**.)

Maasvlakte LNG terminal

Maasvlakte is on the Dutch coast near Rotterdam, one of the world's major hydrocarbon reception and processing areas. To the numerous facilities already at Maasvlakte has recently been added an LNG terminal. One feature of novelty will be described. Heat which might otherwise have been provided by an **SCV** is received from steam at a nearby EON electricity plant. Such steam, which is likely to have been superheated prior to turbine passage, will be saturated on turbine exit possibly with quite a low dryness fraction but will have sufficient enthalpy to be effective in evaporating LNG, eliminating fuel use by a SCV.

McClure Field

In the North Sea, close to **Gryphon**, which, until the recent incident, received from production wells at McClure. The water depth is 115 m and the reservoir at a **TVD** of 163 m. There is a single production well. In development plans for the field, **depletion drive** features.

Macedon Field

Natural gas field offshore Western Australia, currently being developed. It is intended that there will be four production wells, gas from which will sent ashore by pipeline for use in Western Australia and will not (as with so much natural gas from that region) be converted to LNG. (*See* **Angel Field**.)

Machar Field

In the **ETAP**, tied back to the **Marnock Field**. The 16 inch pipe connecting Machar with Marnock is 36 km long and is pigged four times a year. A subsea **pig** launcher was developed expressly for the Machar-Marnock pipeline and it has been in service since 1997. Note that it was an initial design, not the solution to an unexpected problem. Use of a conventional launcher/receiver system accessible above sea level was considered before plans for the innovative subsea launcher were made.

Maersk Gallant

Jack-up drilling rig, having entered service in 1993. It can operate in water depths up to 100 m and can drill to > 7500 m below the sea. Its **drawworks** are rated at 3000 h.p. and its **derrick** can hold a load of 900 tonnes. It is at the time of preparing this entry in service in Norwegian waters at a depth of 73 m. (*See* **Huldra Field**.)

Maersk Guardian

Jack-up vessel in service in the North Sea. Built in 1986 by Hitachi, it can operate in water depths up to 350 m. Its **drawworks** are rated at 2000 h.p. and its **derrick** capacity is 450 tonnes. There are three pumps for the drilling mud. It was used to drill exploration wells at the Olsevar Field, production wells from which are expected to be tied back to the Ula platform. (*See* **Blane Field**.)

MAGNACIDE® 575

Biocide from Baker Petrolite capable of killing **SRB**. It is used in workovers as well as in pipelines.

Majnoon Field

In the south of Iraq, this field is a huge one in which Shell and Petronas each have a holding as does Iraq's own **Missan Oil Company**. It was discovered in 1975. Current production is a modest 45 000 barrels of oil per day, a reflection on the state that Iraq is in generally[40]. The 'Majnoon development contract' is being vigorously implemented by Shell and Petronas. One of the first stages of the development was the award in late 2010 to Halliburton for the driling fifteen wells at the field each having a **pay** of about 200 m. It is believed that this will raise production to 175 000 barrels per day in 2012, almost a four-fold increase in the present figure. The oil field services company Petrofac were in the same year awarded a contract for supply and installation of three gas separation units and other crude processing requisites. The development project is intended to raise the production to 1.8 million barrels per day over a seven-year period even though the 'prewar' level was not above 50 000 million barrels per day. That of course was due to the Iraq-Iran war and its halting of development. Sabotage and looting took place at the field during the war. (*See also* **Az Zubair Field**; **Bazergan Field**; **East Baghdad Field**.)

Malampaya Field

Condensate field offshore the Philippines. The water depth is 43 m, and wells are at subsea depths up to 850 m. A **GBS** is used for production, the support of which necessitated **subsea rock installation** which was carried out by Rollingstone. A quantity of rock totaling 17 000 tonnes was 'installed' to allow for the effects of the unevenness of the seabed, and a further 3000 tonnes were placed at the corners of the GBS. Production facilities at Malampaya can accommodate 14 million cubic metres of gas and more than 30 000 barrels of condensate per day.

Map Ta Phut LNG Regasification Terminal

In Thailand, scheduled for start-up in 2011 with a capacity of five million tonnes per year of LNG. Its features include two large storage tanks with the means of recovering gas having undergone **boil-off**. In order that the terminal could accommodate LNG tankers amongst the largest in use, major dredging work was necessary and for this a **cutter suction dredger** from the Van Oord fleet was deployed.

Mari-B Field

Gas field offshore Israel, discovered in 2000 and producing gas since 2004. The operator is Noble Energy. The sea depth of the field is 243 m and the reservoir depth around 1800 m. The formation is sandstone and the **pay** of gas 168 m. By 2010 there were six production wells yielding more than nine million cubic metres per day of gas. (*See also* **Noa gas field**; **Tamar Field**.)

Markham Field

Gas and condensate field in the southern North Sea straddling the UK and Dutch sectors, discovered in 1984 and producing since 1992. There are two production platforms at the field. The condensate to gas ratio is nine barrels per million cubic feet. A calculation similar to that done for the **Yttergryta Field** whereby this figure is converted to a percentage by weight of condensate is in the shaded area below.

$$9\ \text{bbl} \times 0.159\ \text{m}^3\ \text{bbl}^{-1} \times 700\ \text{kg m}^{-3}/(10^6\ \text{ft}^3 \times 0.028\ \text{m}^3\ \text{ft}^{-3} \times 40\ \text{mol m}^{-3} \times 0.016\ \text{kg mol}^{-1})$$

$$= 0.06\ (6\%)$$

and this makes Markham less abundant in condensate than many 'condensate fields'. Let it be remembered that condensate is a highly valuable product, capable of providing material which can be substituted for petroleum distillates with much simpler refining and no residue. It will be worth the while of the operators at Markham, who include Total and Venture Petroleum, to collect and stabilise that 6%.

Marmara Ereglisi

Scene of an LNG regasification plant in Turkey, 62 miles west of Istanbul. Its nameplate capacity is 4 billion cubic metres per annum, in thermal terms (as calculated in other entries) just under 5 GW. The LNG is sourced from Nigeria and Algeria and some of the gas once 'regasified' is taken to an electricity utility adjacent to the plant.

Marmul to Nimr pipeline

In Oman, of 84 km length and used to convey crude oil. The pipe diameter is 18 inches. It is due for renewal, but plans for replacement have to allow for extension, the details of which are not yet finalised but for which a 24 inch pipeline will probably be more suitable. This would make a newly laid 18 inch pipeline from Marmul to Nimr a compromise from the very beginning of the operation of the extended network. Pending the finalising and approval of the extension plans the operator has achieved extended life on the existing pipeline by use of a **drag reducing agent**.

Marnock

Field within the **ETAP**, producing gas and condensate since 1998. The water depth is 93 m. There are six wells, all of which are almost horizontal. It is unusually rich in condensate. (*See* **Machar Field**.)

Mat support

An alternative to use of the **spud can** in the support of a **jack-up** rig. It consists of a submerged hull, upon which the jack-up rig is supported. The hull is kept in the submerged position by filling its buoyancy chambers with water. Clearly mat support and spud cans have their respective advantages and disadvantages. For example, the latter distributes the weight of the jack-up rig less evenly and leaves a 'footprint'. The former cannot be used where the seabed is sloping or where there are pipelines in the way of laying the hull.

Matrix acidising

Term in well intervention for treatment with acid, sometimes simply called acid treatment or acid stimulation. In a sandstone formation the effect is dissolution of minor ingredients to the enhancement of **porosity** and permeability. In a carbonate formation this itself will of course be attacked by acid. Amounts and concentrations of acid will, in a carbonate application, be adjusted to bring about porosity and permeability improvements in the formation without destroying it.

Matrix body

An alternative to a steel 'body' for drill bits, having **PDC** as the cutting material. The 'matrix' comprises **tungsten carbide** and a copper-nickel alloy. PDC bits have an **IADC** code separate from that applying to steel bits and those made from tungsten carbide only. In the code for PDC bits 'S' as the first character means steel body and 'M' means matrix body. (*See* **Silver Bullet**.)

Maule Field

Oil field in the UK sector of the North Sea, discovered in October 2009 and producing by July 2010 at a rate of 11750 barrels per day. The operator is Apache. The short time between discovery and production was due to the proximity to the Forties field, of which Maule can be seen as a satellite. At present oil from a single production well at Maule goes to infrastructure at Forties. It is expected that a second production well at Maule will be drilled. Another Forties satellite is the Bacchus Field, where development plans are that three production wells will be tied back to the Forties.

MCR®[41]

Multi-component refrigerant, for use in LNG trains. (*See also* **Damietta**; **Yemen**; **LNG production in**.)

MDEA

Methyldiethanolamine. (*See* **ADIP process**.)

MEA

Monoethanolamine. (*See* **Amine process**.)

MEG

Mono ethylene glycol. (*See* **Gel pig**.)

Meged Field

Onshore oil field in Israel only 20 km from Tel Aviv. The operator is Givot Alam. Five exploration wells, Meged 1 to 5 inclusive, were drilled over the period 1994 to 2004 over a field area of 200 square kilometres and to well depths up to 4500 m. It has been stated that a **pay** of 400 m has been identified and that Meged 5 has produced over a short period at about 300 barrels of oil per day. There is a very long way to go before the field becomes commercial[42].

Meji Field

In Nigeria, discovered in 1964. By the early 2000s it had 52 wells, 25 of which were horizontal. The field has several reservoirs and they are separated vertically ('stacked') by formation abounding in **faults**. Chevron Nigeria, using tools developed by Schlumberger, have been installing new production wells at one of the reservoirs already being drawn on, in order to increase production. In accessing the reservoir the drillers were, because of the general non-uniformity and heterogeneity of the formation, very much in *terra incognita*. This gave **LWD**, in particular gamma logging, an important role and a γ-ray detector was placed a short distance above the drill bit. **MWD** including resistivity measurement was concurrent with the gamma logging. The first of the wells was at a **TVD** of 1255 m and the resistivity measurements were decisive in the final stages of its drilling. The measurements indicated an unfavourable change in formation when the drill bit was within four feet of the reservoir boundary, therefore a diversion was made so as to access better rock. During this final stage of the drilling a **fault** with a **throw** of 8 feet was discovered at a measured depth (MD) of 2100 m and the drill had to be steered accordingly. Even from this abstract of what was involved in drilling a well in such difficult circumstances the reader will have appreciated the interdependence of many 'unit ops' (a term from chemical engineering not, to the author's knowledge, usually applied to drilling) including **directional drilling**, MWD in several forms and LWD. Eventually *four* new wells to the reservoir at Meji were drilled. (*See* **Cusiana Field**.)

Mejillones LNG Terminal

In Chile, commencing production in 2010 with a single LNG train capable of producing, from LNG, 5.5 million cubic metres of gas per day. It will be expanded over about the next three years and is intended to ease Chile's reliance on natural gas from Argentina. In fact, such supply has recently been unreliable. At Mejillones, LNG is offloaded from tankers on to a floating storage facility. Further work at the terminal will see at least one permanently installed tank replace this. All of the gas received is diverted to electricity generation. At the present operating level the terminal provides enough gas for 1000 MW of electricity.

Melkøya Island

Off Hammerfest, Norway. (*See* **Snøhvit LNG facility**.)

MER-14X

Recently spudded **CBM** well at the Meridian Field in Queensland, one of nine to have been drilled there. The Meridian Field is west of Gladstone. Mitsui has a 49% interest in the Meridian Field and this reflects Japan's interest in LNG from CBM instead of from conventional natural gas on price grounds. The target drilling depth at MER-14X is 800 metres. Other wells at the field include MER-11V, which has recently undergone **spudding in**. The target depth for that is 1333 m. A previous one - MER-02V - was to a target of only about half that of MER-11V. (*See* **Gladstone LNG project**.)

Mesa 30

Crude oil originating in California, having a density of 30.5 degrees API, density (60 °F) 875 kg m^{-3}. It finds application as a recyclable diluent in the production of particularly heavy crude oil, for example in Venezuela.

Metal Muncher™

Downhole tool from Baker Hughes whose uses include **window** creation. It differs from other devices for the same purpose in that it machines rather than mills. It requires fluid (mud), and this carries away the debris from the machining process. Such debris is of smaller particle size than when a milling approach is taken. Also from Baker Hughes is the simpler Lockomatic™ which *does* use milling to 'intervene' at a casing. The Lockomatic™ can also be used as an **under reaming** tool, in which case it is integral to the drill string.

Methanol, use of in well stimulation

It is mentioned in another entry that the site of **completion** is very easily damaged, and the use of aqueous fluids in well stimulation can indeed lead to damage. In such situations methanol can be used. Before admittance to a well it is made into a foam and a fluorosurfactant is used for this. (*See also* **Blue Angel**; **Formic acid**.)

Miskar Field

Gas field offshore Tunisia, water depth 62 m. There is a production platform operated by British Gas Tunisia, and the gas is rich in condensate. This is taken to the Hannibal Gas Processing Plant where, amongst other things, condensate stripping takes place. The gas so processed will be supplied in quantities of 3.4 million cubic metres per day to a 500 MW power plant. The gas supply and electricity generation are checked for consistency in the box below.

$$[3.4 \times 10^6\ \mathrm{m}^3 \times 37 \times 10^6\ \mathrm{J\ m}^{-3} \times 0.35/(24 \times 3600)\ \mathrm{s}]\ \mathrm{W} = 510\ \mathrm{MW}$$

and agreement is almost exact.

Mizushima LNG import terminal

In south-western Japan, having been in service since 2006, receiving LNG from Oman and from Australia. Expansion is underway so that the facility can start to receive yet more LNG from Australia's North West Shelf. Regasification is by seawater alone by use of **ORVs** or by successive treatment with propane and seawater as at **GDF Suez Neptune**. There is also **SCV** capability and, as a back-up, a heat exchanger by means of which steam can be used to evaporate the LNG. The terminal is at the same site as an oil refinery belonging to the same company. Seawater cooled by LNG regasification goes to the refinery and is used as a cooling fluid ('brine') there. Conventionally brine is cooled by refrigeration[43], so there is an energy saving when the seawater is thus returned from the LNG terminal to the refinery[44]. (*See* **Boil-off**.)

MOAB

Mobile offshore application barge. (*See* **Emeraude Field**.)

Molecular sieve, use of in natural gas dehydration

An alternative to glycol dehydration of natural gas is use of a molecular sieve, by means of which the pipeline standard of 7 lb of water per million cubic feet of gas can be obtained. Such sieves are often composed of zeolites, an example being **MOLSIV™** which can take natural gas down to that value from saturated steam at up to 200 °F (93 °C) in a total (methane + water vapour) pressure range of 100 p.s.i.g. to 1500 p.s.i.g. (788 kN m^{-2} to 10407 kN m^{-2} absolute). A related calculation follows.

Saturated vapour pressure of water at 93 °C = 0.8 bar = 80.8 kN m^{-2}

At the highest pressure in the operating range of the sieve, moles per unit volume =

$$10407 \times 10^3\ \mathrm{N\ m}^{-2}/(8.314\ \mathrm{J\ K}^{-1}\ \mathrm{mol}^{-1} \times 366\ \mathrm{K}) = 3420\ \mathrm{mol\ m}^{-3}$$

$$\text{Amount of water per unit volume} = 3420 \times 80.8/10407 = 27\ \mathrm{mol\ m}^{-3}$$

27 mol m^{-3} of water is equivalent to:

$$27\ \mathrm{mol\ m}^{-3} \times 0.028\ \mathrm{ft}^{-3}/\mathrm{m}^{-3} \times 0.018\ \mathrm{kg\ mol}^{-1} \times 2.205\ \mathrm{lb\ kg}^{-1} \times 10^6\ \mathrm{lb}$$ per million cubic feet

= 30 000 lb per million cubic feet

> The performance of the **MOLSIV**™ therefore is such that it can remove 99.98% of the water initially present.

MOLSIV™ can be used to remove contaminants of natural gas other than water vapour. (*See* **Bettis-DD**.)

Molikpaq

GBS platform, the first such in Russian waters and part of a production area known as Vityaz. Oil from Molikpaq goes to an **FSO** with SPM called Okha, built by Daewoo. This is *not* an **OIB** and is therefore in service at Molikpaq only over July to December, at which stage it is disconnected and towed away. It can, by reason of its strong shell, operate in 'light ice' conditions. Molikpaq is set up for drilling as well as for production. The production capacity is up to 90 000 barrels of oil per day and 2 million cubic metres of gas.

MOLSIV™

(*See* **Molecular sieve, use of in natural gas dehydration**.)

Monopod platform

(*See* **Cook inlet**.)

Moonpool

The part of the hull of a drill ship or a **semi-submersible rig** which can be opened to allow immersion of the drilling assembly. (*See also* **Amirante**; **Discoverer Americas**; **Deepwater Discovery**; **Noble Amos Runner**; **Noble Leo Segerius**; **Ocean Guardian**; **Rollingstone**; **Rubicon Offshore, vessels operated by**; **Seahorse**; **Sedneth 701**; **Shallow seawater, possible difficulties with**; **Side dumping**; **Well Enhancer**.)

MOPU

Mobile offshore production unit. Although as is so often the case some confusion of usage is evident, the distinction between an MOPU and an **FPSO** is clear and simple. An FPSO receives oil directly from the production wells via risers. An MOPU receives oil via a wellhead platform and transfers it on, most probably to an **FSO**. Being free of riser attachment an MOPU is obviously more easily relocatable than an FPSO. Sometimes when a new field is being fast-tracked into production an MOPU is used for production for a limited period whilst further production wells are drilled, and will itself have a back-up role in the drilling. (*See also* **Cendor Field**; **MOPSU™**; **Okwok Field**.)

MOPSU™

MOPU with the additional function of storage on the scale of about 200 000 barrels, possibly eliminating the need for connection to an **FSO**.

Morgan City LA

The scene of inland water exploration for oil (one of many such in the US) where in 2010 there was overturning of a drilling rig. A containment boom for leaked hydrocarbon was deployed and there were no injuries.

Mud acid

Term for a mixture of hydrochloric and hydrofluoric acids when used for removal of filter cake. (*See* **Las Cienegas Field**.)

Munin

FPSO operated by Bluewater (HQ in the Netherlands), to whose fleet **Glas Dowr** and **Uisge Gorm** also belong. Having **dynamic positioning**, it has recently returned to the Lufeng Field after a period of service in the Xijiang Field in the South China Sea. At Lufeng it was involved in sidetracking within existing wells at a sea depth of 330 m. The scene of its current operation is only 250 km from the Hong Kong coast. In the expansion of Lufeng, flexible pipeline from Technip will feature. (*See* **Acergy Condor**.)

MWD

Measurement while drilling, installation of measuring instruments in a well being drilled. Pressure and temperature can be so measured. The drill bit trajectory can be determined and this is especially valuable in **directional drilling**. Instruments and devices for **MWD** are many, and information they provide includes, in addition to what has been noted above, drilled volume, drill bit inclination and occurrence of **lost circulation**. The instruments usually work by magnetic responses and proximity of the instrument to a well or sidetrack other than the one to which the measurements relate can cause interference. MWD tools require electrical power, and a common way of providing this is with a 'mud turbine'. This works simply by diversion of some of the **drilling fluid** into a turbine. It has the advantage over batteries that it can operate at the temperatures of 200 °C or more of higher at which MWD takes place. Nevertheless, there are batteries, notably the lithium thionyl chloride battery, which have been developed for MWD either as the main source or as back-up. (*See also* **AziTrack™**; **DDS™**; **Odoptu project**; **PWD**; **Sorochinsko-Nikolskoe (S-N) Field**; **Sureshot™**.)

MX

Range of drill bits from Hughes with tricone configuration, steel teeth and a **tungsten carbide** insert. (*See* **F21 Well (BP)**.)

My-Lo-Jel

Potato starch preparation from Alliance Drilling Fluids for **filtration** control in water based drilling muds. From the same manufacturer comes Starpac, also for filtration control and containing polysaccharides.

Myanmar-China pipeline project

Work on this began late 2010. It will comprise two pipelines, one for oil and one for gas. Oil will be taken to China from the Middle East and Africa via Myanmar, and completion is expected in 2013. Oil will be initially received at the port of Kyaukpyu in western Myanmar, which can accommodate tankers up to a dead weight of 300 000 tonnes which translates roughly to 2 million barrels of crude oil. Oil enters the pipeline at Kyaukpyu and is taken north-east a distance of 793 km to the border with China, and beyond there in the part of the pipeline within China's borders. The gas pipeline, which has a length within Maynmar of 771 km, will carry gas from the **Shwe Field**. The oil pipeline can deliver at up to 442 000 barrels per day and the gas pipeline at up to 30 million cubic metres per day. The figure for oil is compared with those for the **Kazakhstan-China oil pipeline** and for the **Central Asia-China gas pipeline** below, having regard to the fact that all three deliver oil to China.

Myanmar-China pipeline project	442 000 barrels of oil per day 30 million cubic metres of gas per day
Kazakhstan-China oil pipeline	19 000 barrels of oil per day
Central Asia-China gas pipeline	100 million cubic metres of gas per day
	Total: 461 100 barrels of oil per day 130 million cubic metres of gas per day

The daily oil consumption of China fluctuates more than that of most countries, but 7 to 8 million barrels per day is a reasonable figure from which it can be deduced that the two oil-bearing pipelines in the table will, when the Myanmar-China pipeline is fully operational, provide about 6% of the total. In one sense comparisons between the **Myanmar-China pipeline project** and the Kazakhstan-China oil pipeline are unsound: the former provides a shortcut in shipping for oil produced in distant countries whereas the latter carries oil from source. That is obviously the origin of the factor of about 20 between their delivery rates. It is seen as being strategic in that the pipeline passes through only one country on its way to China and that it prevents tankers on their way to China from having to enter waters known to be threatened by piracy. Ownership of the pipeline will be: China National Petroleum Corporation 50.9% and Myanmar Oil & Gas Enterprise 49.1%. In reviewing the total *gas* supply in the final row we conclude that it is equivalent to being one order of magnitude higher than that from the **Rudong LNG terminal**, which is also in China.

N

Nam Con Son pipeline

This conveys gas from the Lan Tay Field offshore Vietnam to an onshore terminal. Gas from Lan Tay is first diverted to the **Rong Doi Field** for condensate stripping. Condensate is taken to an **FSO** with a capacity in excess of one million cubic metres. Gas from both fields is taken ashore by Nam Con Son. The pipeline is of 26 inch diameter and length 362 km and includes a blanked off **wye piece**. This requires regular pigging to remove the condensate that has deposited onto the pipeline interior surface. The original developers were in favour of the use of pigs in the form of spheres[45]. A point which had to be addressed was that the spheres might not pass easily through the wye piece, a well known effect in flow assurance. In the final design and construction the wye piece was made of transparent glass-reinforced plastic (GRP). This was manufactured in Norway, the development work having been done at Kårstø. Shell had significant activity in these Vietnamese waters until it divested some of its holdings in late 2010 following the Gulf Coast spill. (*See* **Su Tu Den**.)

Nang Nuan Field

Marginal oil field offshore Thailand, water depth 35 m. At peak production the field yielded 12 000 barrels per day. Cumulative production by 1997 was 4.25 million barrels at which stage production was stopped for the time being. This was partly because of difficulties with produced water, which was tending to flash evaporate. Oil production at Nang Nuan was resumed after a **jack-up** rig was put in place there in 2005. Built by Expro, the jack-up rig can receive 12 000 barrels per day of oil and a significantly larger amount of produced water. At present this is taken elsewhere for cleansing, but the jack-up rig can, if required, accommodate a module whereby the cleansing is on site.

Nanhai II

Semi-submersible rig owned by China Oil Field Services, having recently seen duty at the **Shwe Field**, working in 150 metres of water. It is an elderly vessel, having been built in Norway in 1974. It can operate in water depths up to 300 m and can drill to depths of 7500 m.

Natural gas, as a solute in crude oil

When (as is often the case) oil and gas co-exist in a reservoir such co-existence has a number of effects relevant to production, one of which is that, at the pressures of gas in a reservoir, there is phase equilibrium with significant amounts of gas dissolved in the crude oil. It can happen that once a well is declining in oil production terms the gas dissolved in it is of as much value as the oil, in which case depressurisation of the oil field is necessary. The Brent Field in the UK sector of the North Sea began oil and gas production in the 1970s. By 1992 oil production was well down and it was decided to depressurise to get the dissolved gas out. It was estimated that 42 billion cubic metres of natural gas could be so realised. Depressurisation was achieved by continuing with oil and gas production as previously but without the **water injection** which had previously been used to maintain well pressure. The process is gradual, and over the period 1998 to 2010 the pressure of gas dropped from 3500 p.s.i. to 1000 p.s.i. with an enhancement of the gas-to-oil ratio as intended. Gas production is as projected and there is condensate as a bonus. Field depressurisation had never previously been done on that scale.

NBR®

Near Bit Reamer, downhole tool from Halliburton[46]. It is used in **directional drilling** and placed between the **downhole motor** and the drill bit, being compatible with fixed cone drill bits and with roller cone drill bits. It enlarges borehole diameter and as a bonus reduces vibrations at the bit. (*See* **Roller reamer**.)

Nepal Oil and Gas Services Limited

Set up in 2004 to provide Nepal with imported petroleum products. The company states on its web page that it supplies the following: AGO, PMS, DPK and LPFO, acronyms not as widely used as they once were. They will be examined in turn in this entry. AGO means automotive gas oil which is simply diesel. Gasoline would not be classified as AGO because it has never found application as a gas oil, that is, as a feedstock for making synthesis gas or water gas. PMS – premium motor spirit – is high-octane gasoline and is a term which in advanced countries belongs to the early post-war period at the time when car ownership was first becoming widespread. DPK means dual purpose kerosene. 'Dual' means for tractor use and illuminating use: the latter would hardly apply in a developed country. LPFO is low pour fuel oil. 'Low pour' means low pour point, the pour point being the temperature at and below which pouring is precluded by the deposition of wax and asphaltenes which clearly makes for difficulty in usage. These products are of course received as such by Nepal Oil and Gas Services who do not themselves take crude oil for refining. One of the suppliers of the company is North West Oil and Gas Company, HQ in Nigeria. North West Oil and Gas do not themselves refine, but purchase refined product for distribution via a free trade zone at Calabar where they have storage facilities for 0.27 million barrels of refined stock.

New South Wales, proposals for CBM in

New South Wales, like its neighbouring state to the north, Queensland, is active on the **CBM** front. There are of course huge amounts of higher rank coal in NSW. Exploratory drilling for CBM is taking place at the time of writing under the direction of Dart Energy. Estimates of how much CBM there is have been made and vary widely, but are of doubtful importance at this stage as there can be no question that amounts very high by international standards are in place. Plans, which have a long way to go before they are fully implemented, are for an LNG plant in Newcastle NSW, North of Sydney and close to bituminous coal fields. By arrangements reminiscent of the modus operandi of **Pioneer Knutsen**, LNG originating at Newcastle will under these proposals be taken to terminals at the NSW 'central coast' (significantly north of Sydney), at Sydney and at the NSW south coast. This will of course necessitate a terminal of suitable size at each scene of delivery.

Niger Delta, SRB in

A recent study of water in close proximity to oil production facilities in the Niger Delta revealed significant levels of **SRB**. There was a very strong dependence on the concentration of the SRB with season, the highest level being observed during July-August when the rain is most severe. A threat to the sweetness of the oil from the fields was perceived.

Nikaitchuq Field

Oil field in Alaska, producing and also being expanded. It is operated by Eni. The water depth is very low (3 m), although **TVD** values of the production wells are about 1200 m. The field when fully developed will have 52 wells, 22 of which will terminate onshore. This will of course involve **extended reach drilling** and for a production well this means that the site of **completion** is not subsea. Nearly half of the wells are for **water injection**. Target production once the field is developed to this degree is 40 000 barrels of oil per day.

Nile Delta, gas fields in

Producing fields in the Nile Delta include Abu Madi, whose claim to be the biggest gas field in Egypt has never been contested. Discovered 1967, it now has twenty-one wells at around 3000 m depth. Production is 11 million cubic metres of gas per day plus 7000 barrels per day of C_{5+} condensate and 200 tonnes per day of LPG. There are several extensions of the field both on- and offshore. One such is Al-Qar'a discovered in 1985 and producing since 1992. Production there is 5 million cubic metres of gas per day plus 3700 barrels per day of C_{5+} condensate and 200 tonnes per day of LPG[47]. Reservoir depths are around 4000 m. The Port Fu'ad gas field is offshore and is connected by pipeline to gas processing facilities at Port Said. Some of the gas from this facility is sent south east by pipeline to Sinai[48]. The pipeline has a capacity in excess of 10 million cubic metres per day. Another major field in the Delta is Wakar, close to the Port Fu'ad field. These are as stated producing fields. There is much exploration work in the Nile Delta, for example at **Papyrus-1X**.

Nini and Cecilie fields

In the Danish sector of the North Sea producing oil since 2003. The fields are linked by pipeline to each other and to the Siri platform, which is a **jack-up** structure in 60 m of water. The reservoir depth at the Nini and Cecile fields is approximately 2300 m. Gas separation takes place at Siri, and the gas is reinjected to maintain reservoir pressure. A storage tank underneath the platform can hold 50 000 cubic metres (0.31 million barrels). There was cessation of production at Siri between October 2009 and January 2010 when this tank displayed cracks. A drilling rig was brought to the scene and via its **derrick** it provided support additional to that of the caisson, which had become damaged.

Niobrara chalk formation

In WY, the scene of thousands of small gas wells of modest production. Recently LiteProp™, a proprietary **proppant**, has been used there in well stimulation to good effect. (*See* **Nitrogen, as a carrier for proppants**.)

Nitrate ions, action on by SRB

It is explained in the entry on **SRB** how sulphate ions provide for respiration. SRB can also use nitrate ions to respire, according to:

$$NO_3^- + 6H^+ + 5e \rightarrow \tfrac{1}{2}N_2 + 3H_2O$$

Whereas with sulphate the reduced product is either H_2S (poisonous and having a souring effect) or S^{2-} (an agent in incipient corrosion), with nitrate the product is elemental nitrogen which is of course totally harmless. Hence one way of controlling **SRB** is to introduce some nitrate: one nitrate ion so reacted is one sulphate ion not having reacted to form unwanted products. In general when both are present both will be used by SRB and the evidence from laboratory tests is that the greater the proportion of NO_3^- to SO_4^{2-} the better in terms of diverting the SRB from sulphate reduction. A microbial species capable of either metabolic pathway might equally appropriately be called a nitrate reducing bacterium (NRB). When nitrate is introduced into an oil well to achieve this effect it is sometimes seen as stimulating the NRB rather than as suppressing the SRB though surely both are equally true. This has been the strategy for SRB control at fields including **Bonga**, where from the commencement of production there was **water injection**. Nitr*ite* compounds can also compete with sulphate in SRB metabolism, in which case the reaction is:

$$NO_2^- + 6e + 7H^+ \rightarrow NH_3 + 2H_2O$$

Some souring due to SRB has occurred at the **Ekofisk Field** and nitrite has been used to address it. Nitr*ate* treatment has been used at several fields including the **Statfjord Field**. Each is (at least partly) in the Norwegian sector of the North Sea, where major seasonal variations in the sulphate concentration of seawater occur. SRB can also be present in produced water. Recent laboratory tests on produced water from the Lost Hills field gave positive results with both nitrate and **glutaraldehyde**. The latter is simply a biocide whilst the former provides an alternative metabolic pathway as discussed.

Nitrile rubber

Common choice of material for O-rings[49] in well casing. Viton® is also so used, as is neoprene. Prior to the availability of such materials copper gaskets had to be used at intervals along a casing. Nitrile rubber also finds a role in the **annular blowout preventer**.

Nitrogen, as a carrier for proppants

Instead of the fracture fluid, gaseous nitrogen can be used to carry **proppant** into a well, as was the case at the **Niobrara chalk formation**. The figures apropos of that are examined below.

Admitted to 'stimulate' the well: 4500 kg of LiteProp™ slurry, 17 Mm^{-3} of carrier nitrogen

Weight of nitrogen $= 17 \times 10^6\ m^3 \times 40\ mol\ m^{-3} \times 0.028\ kg\ mol^{-1} \times 10^{-3}\ t\ kg^{-1}$

$= 19\,040$ t > 4000 times the proppant slurry weight.

It is mentioned in the discussion of well stimulation that afterwards it is necessary to remove proppant not having been retained in the fracture. With a gaseous instead of a liquid carrier most of the excess proppant is removed with the efflux carrier gas. At offshore installations clean-up involves expensive vessel time, so the savings are considerable if it can be eliminated by use of nitrogen. (*See also* **Fracture pressure**; **Greater Green River WY**.)

Nitrogen rejection

Nitrogen in natural gas is an unwanted diluent having the obvious effect of reducing the calorific value. Unlike carbon dioxide, nitrogen cannot be removed from natural gas by a simple chemical operation, so if separation is required it has to be by distillation. The normal boiling points of methane and nitrogen are respectively 112 K and 77 K so the process is known as *cryogenic* removal of nitrogen. It has been applied to natural gas 35% in nitrogen or even higher (see below). Costain, based in Maidenhead UK, are specialists in cryogenic nitrogen removal from natural gas and the information in the table below is taken from their web pages.

Location of nitrogen rejection by Costain (Date)	Production rate of natural gas. Its initial nitrogen content.
Poland (1977)	7.0 million cubic metres per day. 45% molar basis.
USA (1982)	0.6 million cubic metres per day. 'Variable'
UK (1983)	6.7 million cubic metres per day. 8% molar basis.
Far East (1991)	1.1 million cubic metres per day. 25% molar basis.
Tunisia (1994)	2.0 million cubic metres per day. 55% molar basis.
UK (1997)	5.6 million cubic metres per day. 15% molar basis.
UK (2000)	8.1 million cubic metres per day. 8% to 19% molar basis.
Pakistan (2003)	8.1 million cubic metres per day. 18% molar basis.

Location of nitrogen rejection by Costain (Date)	Production rate of natural gas. Its initial nitrogen content.
UK (2004)	8.4 million cubic metres per day. 12% molar basis.
Pakistan (2005)	4.2 million cubic metres per day. 21% molar basis.
Libya (2005)	4.9 million cubic metres per day. 26% molar basis.
Australia (2008)	11.2 million cubic metres per day. 16% molar basis.
Mexico (2008)	17.6 million cubic metres per day. 15% to 19% molar basis.

Note that when at an oil well nitrogen is injected to increase production it can have the effect of diluting the associated gas. If this is to be collected instead of being reinjected it will need 'nitrogen rejection'. This is in fact done at the Jay Field in Florida, where it was only after a considerable time of nitrogen injection to raise oil production that the associated gas started to show dilution effects. A nitrogen rejection unit was installed to restore the gas to saleable composition. (*See* **Hawkins Field**.)

Njord Field

In the Norwegian sector of the North Sea, sea depth 330 m, productive of oil, gas and NGL from reservoir depths of about 2850 m. There are **faults** in the structure which cause parts containing **pay** to be separated from each other. One Society of Petroleum Engineers document[50] describes Njord as being 'densely faulted' and also informs a reader that **TTRD** has been applied there. A **semi-submersible rig** at the field is used as a floating production platform ('Njord A'). Oil is transferred to tankers, and gas is taken via the **Åsgard pipeline** to Kårstø. The field is in decline and new wells are proposed there which will involve an **AHD** (step-out) of five miles. This is targeted at parts of the formation containing condensate.

NK-12c

Well in the Niakuk Field offshore Alaska. In 2004 a record for **coiled tubing drilling** was set there by Schlumberger. At a measured depth (MD) of 5342 m into an existing well, a sidetrack of 388 m was installed. The **TVD** at its point of termination was 2823 m.

Noa gas field

Straddling the maritime border between Israel and Palestine. The recent explosion at a pipeline in Sinai recorded in an endnote in this volume has stimulated interest in Noa on the part of Israel, so that reliance on supplies from Egypt might be less. Commencement of production from the **Tamar Field** will ease such reliance but in the interim Israel is seeking an agreement whereby it obtains gas from Noa. Proximity to existing infrastructure is such that gas production could commence not more than one year after such an agreement had been made.

Noble Al White

Jack-up vessel, currently operating in Dutch waters. Built in 1982 in France, it can operate at water depths up to 350 m. Its **drawworks** are rated at 2000 h.p. and it has three pumps for **drilling fluid**.

Noble Amos Runner

Semi-submersible rig owned by the Noble Corporation. It is moored by chain and wire, can operate in water depths up to 2400 m, and can drill to 10 000 m. Its **moonpool** is 25 feet by 45 feet and its **drawworks** are rated at 3000 h.p. Its **derrick** is of height 52 m and can hold 680 tonnes. (*See* **Borgland Dolphin**.)

Noble Leo Segerius

Drill ship owned by the Noble Corporation, having entered service in 1981, with rebuilds in 1997 and 2002. It can operate in 1700 m of water and can drill to 6000 m. Its **drawworks** are rated at 3000 h.p. and its **derrick** is of height 56 m. Its **moonpool** is 23 feet by 27 feet and it has **dynamic positioning**.

Noble Phoenix

Drill ship having been in service since 1979. It has **dynamic positioning** and can operate in water depths up to 1500 m and can drill to 7500 m. Its **derrick** can support a load of up to 600 tonnes and its **drawworks** are rated at 7300 h.p. Recently it has been involved in exploration at Geronggong offshore Brunei, in 1000 m of water. (*See also* **Lumut**; **Rasau Field**.)

Noble Roger Eason

Drill ship built by Mitsubishi, entering service in 1977. It can operate in water depths up to 2195 m and drill to well depths of 6000 m. The **drawworks** are rated at 3000 h.p. and the **derrick** can take a load of 630 tonnes. It is a competitive rig. It has received (in 2007) repair by SPS Overlay. It is

Noble Leo Segerius
(courtesy: Cesar Neves)

at the time of going to press recorded as being in Brazilian waters. (See **Conkouati**.)

Noble Tommy Craighead

Jack-up rig, whose tasks have included the drilling of the **Sapele well**. It is capable of operating in water depths in the range 14 m to 300 m, and is supported by **spud cans** of diameter 12 m. It can drill to subsea depths up to 7600 m and its **drawworks** are rated at 2000 h.p.

Non-Newtonian fluid

Products of importance in oil production including drilling muds are such. There are several types of non-Newtonian fluid. One is the Bingham fluid (a.k.a. a Bingham plastic) and most drilling muds including those used for **spudding in** are Bingham fluids. Some formal fluid mechanics is required as background to this discussion and this follows below.

For a Newtonian fluid:

shear stress (unit Pa) = viscosity (units Pa s) × shear rate (units s^{-1})

The shear rate is therefore proportional to the shear stress and the proportionality constant is the dynamic viscosity.

For a Bingham fluid:

shear stress (unit Pa)
= yield point (unit Pa) + [viscosity (units Pa s) × shear rate (units s^{-1})]

When for a Newtonian fluid the shear stress is plotted against the shear rate the result is a straight line passing through the origin. The significance of that is that any non-zero stress will initiate flow. The slope of the line will be the dynamic viscosity of the fluid.

When for a Bingham fluid the shear stress is plotted against the shear rate the plot does not pass through the origin but intersects the vertical axis at a value of the stress called the yield point (YP). Unless the stress exceeds that corresponding to the yield point there will be no flow. The slope of such a plot will be in units Pa s and is known as the plastic viscosity (PV).

The author has examined a source in which such information is reported for a water based mud containing **bentonite** and varying amounts of polyvinyl alcohol. For freshly prepared samples the PV values range from 5 cP to 7 cP (0.005 Pa s to 0.007 Pa s). YP values ranged from

18 lb/100 ft^2 to 39 lb/100 ft^2 (8.7 Pa to 18.9 Pa). A technical bulletin issued by the **drilling fluid** manufacturer QMAX gives a range of 8 Pa to 15 Pa. (*See* **Lumut**.)

Nordnes

DPFV bearing the flag of the Netherlands. Its fall pipe is made of flexible material and has a diameter of 1.1 m. It was built to install rocks at water depths up to 1200 m but has not as yet been used at such depths. It is in fact claimed that the record for rock installation depth was set by Ternes, a DPFV from the same 'stable' as Nordnes, at 887 m. This was in the Mediterranean in 2004.

Norne

FPSO with **turret mooring** in use offshore Norway in an oil field also called Norne at a water depth of 380 m. The turret is attached by 12 mooring lines to the sea-floor and once the 'P' and the 'S' functions - respectively production and storage - are completed the 'O' function (offloading) is on to **shuttle tankers**. Associated gas is reinjected into the reservoir and there is also a well for **water injection**. (*See* **Gjøa Field**.)

North Adriatic LNG Terminal

Off north east Italy, a **GBS** standing in a water depth of 30 m with facilities for regasification. The offshore location necessitates mooring systems for arriving tankers. It receives most of its LNG from Qatar, but some from Egypt, Trinidad & Tobago, Equatorial Guinea and Norway. Its capacity is an annual production of 4.7 megatonnes of gas from LNG. (*See* **Boil-off**.)

North Baja pipeline

This crosses a national boundary, at Arizona (USA) and Baja California (Mexico), carrying gas originating in California and Arizona to Baja California where it is used in power generation. It has one **compressor station** - on the US side - which is rated at 21 084 h.p. (15.7 MW) and gas passage is at up to 14 million cubic metres per day. (*See* **Costa Azul LNG terminal**.)

North Sea, first hydraulic fracture operation in

This was in the West Sole field in the 1960s, prior to production. Crushed walnuts were used as a **proppant**.

North seeking gyro

In oil well drilling amongst other applications a magnetic compass is not suitable for determination of orientation because of possible interference due to stray magnetic fields. When a gyroscope is used as an index of orientation no such interference is possible. The orientation of a spinning gyroscope is independent of any external influences and can be set initially at true north. It is then termed a north seeking gyro and is suitable as a reference in oil well drilling to determine the orientation. Obviously knowledge of the orientation during drilling is particularly important in **directional drilling**: for example, loss of direction might cause accidental intersection of another well with all the danger that that would create. Accordingly orientation determination by gyroscopic means is part of **MWD**. A north seeking gyro device is likely to be used in **whipstock** positioning.

Northern Endeavour

FPSO in long-term service in the Timor Sea on behalf of Woodside Petroleum. It can produce 180 000 barrels of oil per day and can store up to 1.4 million barrels. It has **turret mooring** comprising nine chain-and-wire lines. Design and choice of materials are such that the hull of the vessel has a life expectancy of sixty years. The topside will most likely be rebuilt after twenty years at most. (*See* **Laminaria Field**.)

Northern Producer

FPU, a rebuild having previously been a **semi-submersible** rig. It is in service at the Galley Field in the North Sea which has been producing since 2007. The operator is Talisman. The water depth is 140 m and oil production 6000 barrels per day.

Novolac resin

A.k.a. shell resin, phenol-formaldehyde resin used to coat **proppants**.

NPT

Non-productive time, for example periods during the drilling of a well when the drill is not operating. **MWD** is in some degree directed at reducing NPT.

Nu-Well® 220

Well rehabilitation product from JOHNSONscreens® in Minnesota. It is a polymer dispersant and removes debris from **drilling fluid** and natural clay amongst other contaminants. (*See* **BMR™**.)

Ocean Courage

Semi-submersible rig built at Jurong Island, owned by Diamond Offshore. Its name at the completion of building in 2009 was Petrorig I and it was destined for the Norwegian company Petro Mena. It can operate at water depths up to 3300 m and can drill to a depth of 12 000 m. It has **dynamic positioning**. At the time of preparing this entry Ocean Courage is operating in Brazilian waters and is bearing the flag of the Marshall Islands. Brazil itself negotiated for acquisition of Petrorig I when it came on the market.

Ocean Guardian

Semi-submersible rig having recently seen service off the Falkland Islands. Built in Scotland in 1985 and owned by Diamond Offshore, it has a composite structure **BOP** in which both the ram-type and the **annular BOP** feature. Its **moonpool** is 28 feet by 136 feet. It was taken to the Falklands from Cromarty Firth in Scotland in 2009 by the towing vessel Maersk Traveller in a journey lasting approximately 2 months. (*See* **Shelley Field**.)

Ocean Victory

Semi-submersible rig having seen service both in the North Sea and in the Gulf of Mexico. In 1999 one polyester mooring line was used together with several conventional steel ones whilst the rig was drilling in 1600 m of water for five weeks in the Gulf of Mexico. After that time the polyester line was examined and found to have retained its original condition. This led to increased interest in polyester for such a purpose and in 2001 the semi-submersible rig Ocean Confidence drilled in 1900 m of water with eight mooring lines all of which were polyester. The difference between 'taut' and 'catenary' in platform and rig support has been explained in the respective entries. It is sometimes said that polyester lines never act in a catenary fashion and are always taut. This might mean that a polyester line will not have a visible sag in it as a metal chain would but is nevertheless not necessarily experiencing high tension. Nomenclature here is not unified at the present time. Conventional anchor chains can be affected by **SRB**

and/or **APB** and this is an obvious plus when use of polyester lines is being considered. (*See also* **Sevan Piranema**; **Shelley Field**; **Telemark Hub**.)

ODC-15™

Base oil for **drilling fluid** manufacture, a Sasol product. It is in the kerosene boiling range, with a flash point of 60 °C to 65 °C and a kinematic viscosity of 1.6 cSt at 38 °C. As important as its role as a base oil is that as an additive in water based muds, when it effects an increase in the **clay yield**. There are several other fluids in the ODC range.

Odoptu project

Offshore Sakhalin Island. In a sidetracking operation there the **whipstock** was set at the unusually high angle of 77 degrees to the casing of the well at a measured depth (MD) of 3486 m along it. Although it was brought to a successful conclusion in less time that had been estimated, the operation did have two major challenges. One was the large angle of inclination of the whipstock. The other was the existence of other sidetracks nearby and the possibility that they would cause magnetic interference in the **MWD** instrumentation. (*See* **Rosneft, selected major upstream assets of**.)

Offshore Freedom

New-build **jack-up** rig constructed by Lamprell in the UAE, delivered in Q1 2009. It is able to operate in up to 107 m of water and to drill to 9000 m. It incorporates a 15 K **blowout preventer** and a 5 K one. Its **spud can** diameter is 46 feet. Its **drawworks** are rated 3000 h.p. A closely similar rig built for the same operator is Ocean Courageous. Its scene of construction was Texas, USA. One has to examine the specification sheets for the two very closely to find any differences. (*See* **West Epsilon**.)

OIB

Offloading ice breaker. (*See* **ATOT**; **Molikpaq**.)

Oil and gas separator, principle of

Having regard to the fact that oil so often occurs with associated gas, their efficient separation is important in oil and gas production. In an oil well significant amounts of methane are dissolved in crude oil. Of course, dissolution of methane in the oil is limited and where there is associated gas hydrocarbon exits a well as two-phase flow. A separator can operate across a wide pressure range up to about 100 bar and in so doing remove methane from crude oil. The other factor upon which the action of a separator depends is even simpler: methane is the least dense of all of the hydrocarbon gases, only just over half as dense as ethane which is the next alkane up from methane. Methane therefore rises within a separator enclosure and in that way naturally 'separates'. Residence time in the separator is an important factor in performance. At an offshore production facility water also has to be separated and this again is largely natural because of its immiscibility with oil and with natural gas. The difficulty is that there will have been some emulsification, and an **emulsion breaker** is required. Emulsion formation is promoted by large proportions of heavy material (asphaltenes, waxes) in a crude oil. It is also possible for natural gas condensate to form an emulsion with water and for asphaltenes to stabilise this. An emulsion breaker will often have to be customised for hydrocarbon inventory from a particular well. Some separators, especially those in subsea use, operate along the principle of a cyclone, as with **CySep™**.

Okha FSO

FSO currently in service at Sakhalin, destined for conversion to an **FPSO** with a **disconnectable turret** for service off Australia's North West Shelf. The conversion, which is by a contractor on a turnkey basis, will use the turret from Cossack Pioneer, an FPSO having served off Australia's North West Shelf that is now to be decommissioned.

Okoro Field

Offshore Nigeria in a water depth of 14 m, producing since 2008. A vertical appraisal well drilled by the **jack-up** rig Seadrill 7 (previously called Toto) to a depth of 1980 m, and a sidetrack well from that, revealed 21 m of **pay**. Another jack-up rig – **GSF Adriatic VI** – was used to drill production wells, two initially and then a further five. Production is by means of an **FPSO** called Armada Persaka. This can hold 0.36 million barrels of crude oil and it is expected to be in service at Okoro for five years. Current production is 16 000 barrels per day. (*See* **Yoho Field**.)

Okwok Field

Oil field offshore Nigeria, discovered in 1967 but only now being developed on a loan-financed basis. Water depths are 35 m to 50 m and the operator is Oriental Energy. **Spudding in** of an appraisal well was in Q3 2010. Nearby (15 km) is the Ebok Field (water depth 41 m) which started producing in Q1 2011 although exploration and appraisal activity continue there. At Ebok, which is operated by Afren, production is from a wellhead platform whence the oil is taken to an **MOPU** and from there to an **FSO**. It has been noted that Okwok and Ebok have different operators, and both are 'marginal fields'. The viability of Okwok depends strongly on arrangements whereby it will be tied back to infrastructure at Ebok. To this end Afren have recently obtained a stake in Okwok. (*See also* **Okoro Field**; **Yoho Field**.)

Olowi Field

Oil and gas field offshore Gabon, water depth 30 m. The first production was in 2009 and the operator is Canadian Natural Resources. The **pay** is 23 m and initial drilling was by **Ben Avon**. Production and development are concurrent as would be expected for a field having entered production only about a year ago. The tanker Knock Allan was converted to an **FPSO** expressly for Olowi and is under contract to operate there for 10 years. Retaining the name Knock Allan, it has spread mooring and can hold a million barrels of crude oil.

Omnithane®

Polyurethane product from Pipeline Engineering (HQ in the UK) uses of which include the **foam pig**.

OOL

Oil offloading line, for transfer of oil from an **FPSO** most probably to an **FSO**. At the **Bonga** field OOL made of TRELLINE™, comprising bonded rubber, is in use. Rubber products had previously been used in OOL fabrication, but steel is more common.

Orisant

Trailing suction hopper dredger based for most of the time in the North Sea. It can operate at sea depths up to 35 m. Like other offshore vessels of its type it has no cutting tool, only a pump to provide suction. It raises up to 10 000 cubic metres of seawater per hour and has a suction pipe of diameter 800 mm. A simple calculation follows.

Taking the Orisant to be working at its maximum depth of 35 m and the hopper to be at **BCD** zero and using a density value for pure water:

$$\text{rate of work} = (10\,000\ \text{m}^3 \times 1000\ \text{kg m}^{-3} \times 9.81\ \text{m s}^{-2} \times 35\ \text{m}/3600\ \text{s})\ \text{W} = 954\ \text{kW}$$

and this is about 15% below the nameplate rating of the pump, which is 1120 kW. This difference is probably due to the higher density of the 'dredge spoil' than of water. A value of 1174 kg m^{-3} for the spoil would have fitted the nameplate capacity. A calculation follows.

Assigning the mineral matter suspended in the dredge spoil the representative density of 2600 kg m^{-3} (actually the particle density of sand) and letting its fraction by weight in the spoil = $\boldsymbol{x}$:

$$1174 = 2600\boldsymbol{x} + 1000(1 - \boldsymbol{x})$$

$$\downarrow$$

$$\boldsymbol{x} = 0.11\ \ (11\%)$$

The feature for which the Orisant is usually noted is its 'self-unloading system' whereby a conveyor enables the hopper contents to be transferred to shore at a rate such that a full hopper can be completely emptied in about two-and-a-half hours.

Orkid Field

Offshore Malaysia and productive of oil and gas. It has recently become the scene of an **FSO** having single point mooring with nine catenary lines secured at the sea-floor by piles going 30 m deep. Converted from a tanker, the FSO can hold up to 0.75 million barrels. (*See* **Gannet Field**.)

Orlan

Platform offshore Sakhalin Island, from which wells at one side of the **Chayvo Field** have been drilled. Oil and gas are taken from the wells by risers to the platform and from there to the onshore facilities at Chayvo beach. All of the oil from there goes to a terminal at Khabarovsk Krai for export. Some of the gas is supplied to the domestic market, although some is reinjected into the reservoir in order to sustain its internal pressure ('reservoir support'). The target oil production at the Orlan platform is 250 000 barrels per day. More wells are proposed than are currently drilled and the platform already has the capacity to receive oil and gas from them. (*See* **Icebreaker Sakhalin**.)

Ormen Lange Field

Major gas field in the Norwegian sector of the North Sea. The water depth is 800 m to 1100 m. Drilling was from two rigs: **West Navigator** and Leiv Eirkisson. The latter is a **semi-submersible rig** having also seen service offshore Angola. There are now six production wells and an ultimate production rate of 70 million cubic metres per day is aimed for. The field has a life expectancy of 40 years, and the production facilities will sooner or later be upgraded. Both the **tension leg platform** and the **spar platform** are being evaluated. Gas goes to the terminal at Hyamna. Ormen Lange is the scene of experimental use of a subsea gas compressor which operates at 12 MW (16 092 h.p.).

ORV

Open Rack Vaporiser. Heat exchanger used to convert LNG to gas with seawater as the warmer fluid. Because of the high temperature difference between the LNG (enters at −160°) and the seawater (enters at 10°C or higher depending on local conditions), mild steel, widely used in heat exchangers for conventional use, is not suitable. Aluminium, a much better conductor of heat than mild steel, is the usual choice. Seawater in the shell of the exchanger will run over tubes with the LNG inside them.

Sometimes one or more ORVs alone are sufficient for base load, and **SCV** will be started up when demand is high. (*See also* **Altimira LNG terminal**; **Hazira LNG terminal**; **Incheon LNG terminal**; **Mizushima LNG import terminal**; **Sodegaura LNG terminal**.)

Oseberg Field

In the North Sea, 85 miles north-west of Bergen, an oil field having a **gas cap**. The sea depth is about 100 m and reservoir depths below the sea are in the range 2300 m to 2700 m. The operator is Statoil. There are four major installations, Oseberg A,B,C and D. Oseberg A is a **GBS**. Oseberg B is a **fixed platform** from which drilling takes place. Oseberg C is set up for drilling and production. It receives oil from eighteen well production wells. There are three wells for **water injection**. Oil and gas go from Oseberg C to Oseberg D for gas processing including **condensate stripping**. Oseberg D was the most recent of the four to enter service. Prior to its existence all of the gas was reinjected into the well. Osberg D has an additional role in gas exploration. Oil production at Oseberg is typically 30 000 barrels per day.

OTTO system

Procedure developed by SubSea Systems in Aberdeen in the late 1980s whereby in **underwater welding** a human diver sets up the equipment for welding - electrode, filler metal, inert gas shield - where it is needed and having done so is able to surface and to operate the equipment from onshore or from a vessel. Techniques for welding from a distance were developed for the nuclear industry where in hands-on work the welder would have been endangered by emissions from radioactive substances. The OTTO method in its development drew on that. OTTO is currently developed to be applicable at sea depths down to 625 m.

Owo field

Prospect offshore Ghana, licensee Tullow Oil plc. The Owo-1 exploration well revealed 53 m of net **pay** of oil. Limited **directional drilling** from the primary well ('sidetracking'), conducted by a **semi-submersible rig** with **dynamic positioning**, demonstrated continuity ('communication') with another oil field, pushing the pay up to 200 m. Below the oil is natural gas, about two thirds of it containing condensate.

P 110

The point was made in the discussion of the **casing exit system** that there had been such an operation with a single trip of the milling assembly at a well where the casing was made of P 110. This has a design stress of 758 MPa, twice that of j-55 ('casing grade') with which P 110 is often compared. (*See* **Casing exit system**.)

Pajarito-IX

Well in the Carbonera Block in Venezuela, recently drilled to a depth of 610 m, that is, a moderate depth only making development more viable. **Pay** zones of 6.5 m and 10 m were identified. The oil so far obtained at the well is of API gravity 41 degrees (density at 60 °F = 820 kg m^{-3}) making it fairly light (in contrast to that from the **Zuata Field**) and this too is an incentive for development. However further exploration before production can be considered will require **pump jack** installation.

Palanca

FSO in service offshore Angola in shallow waters, a tanker converted to an FSO. Capable of holding 1.6 million barrels, it has an **OOL** of 12 inch diameter.

Palygorskite

Synonym for **attapulgite**.

Pañacocha Field

Oil field in Ecuador (an OPEC country). It is in a region occupied by indigenous people and some of the profit from the oil will go to them. Even so, the financial basis of the field development has met with some controversy as social security funding has been diverted to it. China, always seeking more oil, has made a $US1 billion payment to Ecuador in return for an undertaking to supply it with 96 000 barrels of crude oil a day for two years. At the OPEC basket price, which will presumably apply to the oil so supplied to China, the total value will be about $US8 billion, so China is paying something like a 12.5% fee for assured supply. Pañacocha Field is currently producing about 25 000 barrels per day.

Panama canal, passage of LNG carriers through

Expected to be permitted for vessels conforming to size and configuration specifications from 2014. The trading of natural gas between Asian countries and the US will be strongly affected. (*See* **Freeport TX**.)

Papyrus-1X

Well drilled by **Atwood Aurora** offshore the Nile Delta in a water depth of 20 m. The well depth is 2000 m and the formation is composed of sand varying in particle size from coarse to very fine ('turbidite sand'). Papyrus-1X is in the West El Burullus Concession and was preceded by a well of depth 2043 m drilled by the Diamond offshore **jack-up** rig Ocean Spur. Nearby is another gas prospect called Bamboo, to which Atwood Aurora was moved after duty at Papyrus-1X. An option being considered by the developers is **completion** of Papyrus-1X in such a way that it becomes a production well. (*See* **Galoc Field**.)

Pathumabaha

FSO in the Bangkot field offshore Thailand used solely for condensate. A 'new build' (that is, not a conversion as with **Kome Kribi 1**), it stands in 78 m of water and can hold 400 000 barrels. Mooring lines go to an **external turret**, permitting weather vaning.

Patos-Marinza Oil Field

In Albania, discovered in 1928 and producing since the 1930s. About 12 miles from the coast with the Adriatic Sea, the field produces heavy oil, some of it as high as 1050 kg m^{-3} in density[51]. The cumulative production by the middle of the first decade of the 21st Century (that is, about 70 years after production began) was 120 million barrels, justifying the view that the field is a 'major' one. The field covers a large area. Well depths are in the range 60 m to 1800 m, porosities in the range 15% to 40% and permeabilities in the range 1 millidarcy to 2000 millidarcy. That the field started to become moribund over time was, given its location, inevitable, and there has been revival recently. It was reported in April 2011 that in the first quarter of that year Patos-Marinza had been producing at an average rate of 11 894 barrels per day. This can be compared with the average rate over the seventy-year period previously referred to, which would have been:

$$120 \times 10^6/(365 \times 70) \text{ barrels per day} = 4700 \text{ barrels per day}$$

so recent production is over twice that. Also during the first quarter of 2011 sixteen horizontal wells had been drilled, thirteen of which had undergone **completion**. There has also been workover of some of the (numerous) vertical wells already drilled over the period of operation of the field. In April 2011 it was stated that 57% of the oil production was from the new horizontal wells and 43% from the reconditioned vertical ones.

Pay

A.k.a. pay zone, the part of a reservoir from which hydrocarbons can be recovered and a term featuring widely throughout this volume. An important part of the viability of a field under appraisal, pay is determined during exploratory drilling and is expressed in units of depth. When at an exploration well potentially productive zones are separated by non-viable ones, the total depth is the gross pay, and the sum of the axial lengths of the productive zones the net pay. 'Viability' depends not solely on presence of hydrocarbon but also on permeability and **porosity**, so, as such things as hydraulic fracture advance the criteria for viability change. Note that the term 'pay' applies, with the same meaning, to **CBM**.

PDC

Polycrystalline diamond compact, material for drill bit fabrication. It contains both diamond and **tungsten carbide** and is made by sintering. The Mohs hardness can be adjusted across a range of about 5 to 9 by composition and conditions of sintering. A suitable PDC bit can be selected on the basis of the properties of the rock being drilled. For example calcite has a Mohs hardness of 3, and the feldspar and quartz granite, granite 6 to 7. A tool for reaming might also use PDC. (*See also* **Bit Booster®**; **Drag bit**; **Drill bit, life expectancy of ('bit life')**; **F21 Well (BP)**; **Formation structure, hardness of on the Mohs scale**; **FXTM**; **GaugePro XPR™**; **Matrix body**; **PDW 6**; **QuantecTM**; **ROP**; **RwCTM**; **Silver Bullet**; **Spear drill bit**; **Tiber Field**.)

PDQ

'Production, drilling and quarters'. (*See also* **Azeri-Chirag-Gunashli Field**; **Visund Field**.)

PDW 6

Range of drill bits manufactured by Geogem in the UK, suitable for **directional drilling**. The cutting edges are composed of **PDC** with the crown of the bit – the outermost part – made of **tungsten carbide**. The PDW 6 comes in diameters from 3 inch to 6¾ inch.

Perdido platform

A **spar platform** in the Gulf of Mexico, operated by Shell. Rather unconventionally for a spar platform, it has both drilling and production facilities. This requires movement capability of the spar platform from one drilling scene to another, and limited movement is attained by adjustment of tension in the nine mooring lines. **ClearPAC** was specifically developed for use at Perdido. Production began in March 2010. The water depth is 2285 m and the well depth 2400 m. In a novel arrangement, oil from multiple wells is taken to a caisson[52] and from there to the platform by many fewer risers than would be required if there were one per well. The sea-floor area accessible to the platform is increased by this as well as by the movement capability previously described, and this represents an advance over previous operations at such depths. Separation of oil and gas is subsea. This suits the requirements at the platform, but in any case 'downhole separation' is preferable when an **electric submersible pump** is being used as at Perdido, as gas decreases the efficiency of such a pump. The electric submersible pump at Perdido is for oil transfer from the caisson upwards and is rated at 1500 h.p. This point is addressed in the box below.

In the discussion of the **pump jack** it is estimated that at the Lost Hills Field in the US 785 pump jacks each of 0.69 h.p. would produce 33 000 barrels per day.

Peak production at the Perdido platform is 100 000 barrels per day.

If the efficiency of the electric submersible pump is the same as that of the pump jack this requires:

$$(100000/33000) \times 785 \times 0.69 \text{ h.p.} = 1641 \text{ h.p.}$$

and agreement is within 10%.

Implicit in the above calculation is the assumption that the caisson depth at Perdido is the same as a typical well depth at Lost Hills. The well depths at Lost Hills actually range by an order of magnitude. In more routine applications an electric submersible pump is of course in the well casing. (*See* **Genesis Field**.)

Peregrino Field

Offshore Brazil, in 100 m of water, productive of a heavy crude. It is being developed by the Norwegian company Statoil. Horizontal drilling is taking place and it is planned that there will be 30 production wells, oil from which will go to two production platforms and from there to an **FPSO** also called Peregrino. There will also be 7 wells for **water injection**. The FPSO, commissioned specially for Peregrino, will be a rebuild of a vessel previously in tanker service. It will be conventionally moored at Peregrino with chains and mooring lines. Oil from both production platforms will be taken to the FPSO in what is a fairly unusual arrangement. In fact the 'production platforms' in this entry are in some accounts of the Peregrino Field called 'wellhead platforms' on the basis that they convey oil to the FPSO which by definition is the scene of 'production'. There is no reason at all to dissent from this terminology. (*See also* **Jasmine Field**; **SWHP**.)

Perforating gun

An alert reader having studied both the entry for **BKB** and that for **completion** will have perceived a difficulty as follows. The first section of drill tube admitted is that which, if 'conventional perforation completion' is applied, will need to have perforations. The kelly operator will not know how many sections of drill tube will be needed to reach the ultimate termination of the well casing and anyway a drill tube section *admitted* with perforations would make for serious difficulties with casing installation. This means that at completion the casing has to be perforated in situ and a perforating gun is used for this. This contains an explosive capable of creating a hole in the drill tube. It can be admitted to the well by wire or cable. Alternatively it can be admitted by coiled tubing in which case the term **TCP** – tube-conveyed perforating – applies. Obviously the 'gun' diameter has to be less than that of the casing into which it is lowered. The performance of the 'gun' will depend on the number and configuration of explosive charges and can be expressed in 'shots per metre'. As an example, TCP guns from the international oil field services company Weatherford International range from 3 to 20 shots per metre. Weatherford also manufacture the **SureFire™**, a perforating gun descent of which to the site of completion is by wire, i.e., it is not a TCP device. A **slickline** can be used to fire a perforating gun. (*See also* **Lunskoye-A**; **San Jaun Basin NM**.)

Performax™

Water based **drilling fluid** manufactured by Baker Hughes. It has recently found application in Nigerian waters as a replacement for emulsions over which it has an advantage in environmental terms by not containing any organics which can be released into the sea. The Performax™ in this application used seawater as an aqueous base. It also comes in a freshwater version which has recently been used in Colombia.

Perlite

Insulating material suitable for LNG, available in bricks and in loose form.

Its constituents are mainly silica and aluminium oxide and when heated it behaves like glass. In a double skin LNG tank, perlite in loose form will occupy the space in between the layers. Over a time much shorter than the life expectancy of the tank itself, the perlite can, because of the extreme temperature gradients it experiences, shrink. This makes it less effective as a insulator, creating 'cold spots' on the tank. Renewal of perlite in situ is possible in such circumstances by injection of perlite via a 'lance', with nitrogen at high pressure as the carrier fluid. This operation is routine and was recently performed at **Grain LNG Terminal** and at the LNG storage facility at Glenmavis in south-west Scotland.

Pertamina

State owned Indonesian oil company. (*See also* **Barito Basin**; **Cepu Field**.)

Peru LNG Project

This 'project' involves the first LNG plant in South America, dispatching its first shipment to Costa Azul LNG terminal in 2010. The liquefaction plant at Pampa Melchorita comprises a single LNG train. It receives some of its natural gas from the Camisea Gas Field and the remainder from the Chiquintirca gas field in the Andes. Pipelines from each field go to Pampa Melchorita.

Peter Schelte

Pipe laying vessel under construction in Korea, expected to enter service in 2014. The operator will be AllSea, whose fleet also includes **Solitaire** and **Audacia**. Peter Schelte will have **dynamic positioning** and a stinger length of 170 m. It will be capable of laying pipes up to 68 inch o.d.

Petrobras XXI

Semi-submersible rig which whilst on tow from Brazil to Cape Town in 2006 became detached from its tug near Tristan da Cunah. Sight of the rig was lost from some days and it was eventually found to have come ashore at Trypot on the SE of Tristan da Cunha. The eventual effects await evaluation, but there is the point that a semi-submersible rig does not *carry* crude oil so comparisons with oil tankers having experienced a similar fate need not be made.

Petronius A

Platform in the Gulf of Mexico, an example of a compliant tower. It stands in 550 m of water and produces 42 000 barrels of crude oil per day with large amounts of associated gas. It suffered damage as a result of Hurricane Ivan and was out of service for about six months for repairs. All compliant towers were in the Gulf of Mexico until Angola took up the challenge with **Benguela Belize**.

PFW

Produced formation water, water accompanying oil and gas from a reservoir. (*See* **CrudeSep®**.)

PI

Productivity index, units barrels per day per p.s.i. The pressure part of the definition is based on the drawdown pressure, representing the pressure difference between reservoir and well: this is controllable by choking. No distinction is made between on- and offshore fields in the application of the definition of PI. There are serious difficulties in interpreting different PI values across a range of wells. One is that circumstances can be adjusted to raise it: choking was mentioned earlier as such a means of adjustment. Moreover, the features of the **completion** have an effect. A PI value of the order of 1 bbl d^{-1}p.s.i.$^{-1}$ on superficial consideration would not augur well yet such a value applies to the Beryl Field even after well stimulation (see below). Moving from the particular (the Beryl Field) to the general and hypothetical, it only requires a modest pressure value of 500p.s.i. for a PI of unity to become 500 barrels per day. It is shown in the entry for the **pump jack** that at an onshore field such a pump operating at about 1 h.p. would produce of the order of 50 barrels per day, so at an onshore field a PI of unity at a pressure of 500 p.s.i. is equivalent in performance to ten pump jacks. No doubt for an offshore field the equivalence could be estimated on the basis of the **electric submersible pump**. Yet PI values can exceed unity by a couple of orders of magnitude. These sorts of perspectives have to accompany any interpretation of a PI value. Presence or absence of associated gas is relevant for two reasons: its role in the pressure and, if it is collected for use, its effective addition to the 'barrels' in the formulation with regard to what is said in the entry for **Associated gas, value of**. The table below contains some PI values for wells at various fields. They have been included for their interest and must be considered against a background of what has been said in the previous few sentences. Notes follow the table.

Field	**Productivity index/bbl d^{-1} p.s.i.$^{-1}$**
Beryl, North Sea	1.4 (*See* **North Sea, first hydraulic fracture operation in**)
South Brae, North Sea	10 to 40

Field	Productivity index/bbl d^{-1} p.s.i.$^{-1}$
Ghawar, Saudi Arabia	31 to 141 across the different areas of the field.
Round Mountain, California	140
Thirty wells from fields offshore Brazil considered collectively.	65 to 80

The source from which the PI for the Beryl field was taken gives a production rate after well stimulation of 2400 barrels per day, indicating that the pressure was

$$(2400/1.4) \text{ p.s.i.} = 1750 \text{ p.s.i. approx.}$$

The PI at South Brae, also in the North Sea, is much higher than that for Beryl. Oil from this field is accompanied by large amounts of gas. The Ghawar Field (row 3) is the largest oil field in the world in terms of proven reserves. The wells considered in the fields offshore Brazil (final row) range in **TVD** from 380 m to 3000 m. Directional drilling took place at all of them: the smallest **AHD** across the range of 30 wells was 188 m. Each of the 30 wells has a production of the order of one million barrels per year. (See **Tengiz**.)

Piceance Basin

Pronounced 'Peeawance', in north-west Colorado, a source of large amounts of natural gas. Access to the gas is challenging (it is 'tight gas'). Target production over the life of the field at Piceance, which Exxon Mobil are currently operating, is 1.25 trillion cubic metres. There had been gas production at Piceance for many years previously from a conventional reservoir. What is left is in a very large number of small reservoirs enclosed by hard rock in close proximity (typically 5 metres) to each other. These are initially approached using the ExxonMobil Fast-Drill Process. A hydraulic fracturing fluid is then pumped down the casing at sufficient pressure to break open the rocks containing the gas, releasing the contents of previously isolated pockets. The lowest part of the casing is perforated and through it natural gas is able to ascend for collection. The pumps which inject the fluid are mobile, being mounted on trucks. Current production is about 20 million cubic metres per day. Piceance Basin has recently been the scene of very precise and challenging use of a north seeking gyro. A new well was being drilled in close proximity to 16 others which were already at the spudding in stage or further and intersection of the drilling trajectory of the new well with one of the existing wells had to be avoided. At a measured distance (MD) of 1300 m along the well being drilled the inclination was measured as 7.38 degrees. The azimuth – angle with due north therefore being in the range 0° to 360° – when determined on two occasions about ten weeks apart was 212.06° at the first measurement and 212.16° at the second, representing very high precision. (*See also* **GyroTracer™**; **USA, selected CBM reserves in**.)

Pickerill Field

In the North Sea, productive of condensate-free gas since 1992. Until 2003 it was operated by BP, since then by Perenco. The water depth is 21 m. Between May 1991 and May 1994 eight wells were drilled at the field. In 1997 a *multi-lateral well* – one of the first in the North Sea – was drilled. Development continued and three sites of **GWC** were found at **TVDs** in the range 2660 m to 2770 m. There are two unmanned platforms at Pickerill. Their gas is taken to the terminal at Theddlethorpe on the coast of Lincolnshire and received there on a third-party basis by Conoco-Phillips who operate the terminal. (*See also* **Boulton Field**; **Changbei shale natural gas field**.)

Pig[53]

(*See* **Foam pig**; **Gel pig**; **Utility pig**.)

Pioneer Knutsen

LNG carrier which takes LNG on board at **Kollsnes** in small amounts and delivers it to purchasers within Norway. Its range is therefore local. Published specifications for the vessel give a deadweight of 640 tonnes and a cargo tank capacity of 1100 cubic metres. Dividing one by the other gives:

$$640000\ \text{kg}/1100\ \text{m}^3 = 581\ \text{kg m}^{-3}$$

The density of LNG varies with composition, as would be expected, and the value calculated above is on the high side: values in excess of 500 kg m^{-3} are unusual. The discrepancy is due to disregard (through lack of information) of the contribution made to the deadweight additional to that made by cargo. The vessel is of small capacity for a restricted range of operation. Such an LNG carrier is dubbed a **Kyoto tanker**. The very small scale of the Pioneer Knutsen can be appreciated by comparison with LNG tankers supplying, for example, the **Rudong LNG terminal**, payloads of which are in the range 80 000 tonnes to 267 000 tonnes of LNG. (*See* **New South Wales, proposals for CBM in**.)

Pipe laying vessel

(*See* **Acergy Condor**; **Audacia**; **GSP Bigfoot 1**; **Corrib Field**; **Deep Blue**; **J-Lay**; **Leighton Eclipse**; **Lorelay**; **Peter Schelte**; **S-Lay**; **Saipem 7000**; **Solitaire**; **Tog Mor**; **Underwater welding**.)

Pipeline repair clamp

A common means of repairing a damaged pipeline is with the split sleeve clamp. Such a clamp is initially divided into two parts, each having a half-circle space of radius close to the outer radius of the pipeline being repaired and coated with an elastomer. It is installed by bringing the two half-circles together at the site of the damage to the original pipe and bolting them into position so as to repair the pipe. An example is Tekclamp®, manufactured by Tekmar Subsea. It is available for a wide range of pipeline sizes and when installed in the authorised way can withstand pressures up to 8800 p.s.i. (*See also* **Collet-Grip™**; **Everest Field**.)

Piranema Field

Offshore Brazil, water depth in the range 1200 m to 1600 m. Production is by **Sevan Piranema**, which is an **FPSO**. The field is poised for expansion of production and further wells are being drilled at the field by **semi-submersible rig** Ocean Winner, a member of the fleet of Diamond Offshore (HQ in Houston with an office in Macae, Brazil).

Platong II Project

Expansion of an existing gas production facility in the Gulf of Thailand, completion expected 2012. The water depth is 70 m and the expanded production facility will have a nameplate capacity of 12 million cubic metres per day. There will be facilities for drying the gas to give levels of water below which there will be no condensation in the pipeline, and also for condensate stabilisation. There will, in fact, when the platform is working at full capacity, be 22 000 barrels per day of condensate. By means of a calculation like that in the entry for the **Yttergryta Field** it can be shown that this represents 31% by mass of condensate. Produced water exceeds in volume the condensate yield. The condensate goes to an **FSO** called Pattani Spirit, part of the Teekay fleet. This is a tanker conversion with SPM and has a payload of 0.85 million barrels.

PNG LNG project

This is expected to enter production in 2014, producing 6.6 million tonnes of LNG by two LNG trains on a site close to Port Moresby. The operator is Esso Highland and gas will be sourced from the **Hides Field**, the Angore Field and the Juha Field. Condensate removed prior to liquefaction will take place using existing facilities for gas from the Kutubu (Agogo) Field, and will be exported from the **Kumul Marine Terminal**. (*See* **Kutubu Oil Project**.)

POB

Personnel on board an offshore unit. (*See also* **David Tinsley**; **Stena Don** *inter alia*.)

Poland, shale natural gas in

Drilling has begun near Gdansk for shale natural gas, participants being ConocoPhillips, ExxonMobil, Marathon and Talisman. Poland is currently dependent on Russian imports of natural gas from Gazprom. Hydraulic fracturing will take place. The paucity of onshore drilling rigs available for the 'majors' having established a presence at the gas play near Gdansk is seen as a possible factor causing delay. (*See* **Urengoy Field**.)

Polecat Creek well TX

The scene of recent very effective application of acid stimulation. An order-of-magnitude increase in production was achieved. (*See* **Bluell formation ND**.)

Pompeii

Dynamically positioned **SSDV** operated by Tideway. It is an example of an increasing trend whereby a vessel can be used in offshore wind farm installation as well as in oil and gas production. Pompeii is currently being evaluated for use in cable laying at offshore wind farms.

Porosity

The fraction or percentage of the apparent volume of a solid actually occupied by voids. This is a totally general definition applying to anything being relevant, for example, to coal science. In oil and gas production it is relevant to geological formations, also to gravel packs and to **proppants**.

Potassium, Thorium and Uranium in well formations

The isotope Potassium-40 decays to Argon with β emission and γ emission of photon energy of 1.46 MeV, signifying a photon of frequency 3.5 $\times 10^{20}$ Hz and wavelength 8.5×10^{-13} m. Thorium-232 and Uranium-238 undergo a sequence of radioactive processes, the daughter products of which release γ radiation. It is the presence of Potassium-40, Thorium-232 and Uranium-238 in rock that is basis of 'gamma logging' in **LWD/MWD**. The photon energy expected from Potassium-40 is a single one; this is not so for the other two isotopes which as explained above form over a number of steps several products capable of γ emission. The result is that in gamma logging a range ('spectrum') of wavelengths of γ radiation will be encountered by any instrument placed downhole to detect it.

Potato starch

This is used in some water based drilling muds in **filtration** control and/or viscosity adjustment. An example is **My-Lo-Jel**. Corn starch is also so used. Pre-treatment of starch for inclusion in drilling muds is sometimes microbial, as with **DEXTRID®**. (*See* **Aqua-Clear®**.)

ppb

Pounds per barrel. (*See also* **Invermul®**; **Thamama**.)

ppg

Pounds per (US) gallon, unit of density for **drilling fluid**. Fresh water has a value of 8.33 ppg. Values below that for pure water can be obtained by passing air or nitrogen into the fluid to give it a foamy consistency in which case a foaming agent will also be required. (*See also* **Bentonite**; **BMR™**; **Ceiba Field**; **Sidetracking, by coiled tubing drilling**.)

Prime Plus™

Resin coated **proppant** from Hexion Specialty Chemicals in Columbus OH. It is recommended for use at higher closure stresses, that is, between about 6000 p.s.i. and 10 000 p.s.i. In tests in Lower Cotton Valley TX, Prime Plus™ was used in one well and a competitor's product in another, and over a 90-day trial production was 300% higher with Prime Plus™. Inclusion of such information is not to promote Prime Plus™ but to make the point that where there has been hydraulic fracture production has a strong dependence on the nature of the proppant. (*See* **Casing exit system**.)

Prirazlomnoye Field

Offshore Russia, water depth 20 m and 60 km from the shore. The shallow depth makes for ease of installation of a self-supporting platform[54]. Forty wells are proposed: nineteen for production, sixteen for **water injection** and five to be held in reserve after **completion**. Whilst the shallowness is on the side of the field operators, the extremely harsh weather conditions make for difficulty. The area is ice-free for less than a third of the year; wind speeds are up to 40 m s^{-1} and wave heights up to 12 m. Current production – poised for increase – is about 14 000 barrels per day with 10^6 m^3 of associated gas. (*See* **Associated gas, value of**.)

Production rigs, propulsion of

Some offshore production rigs are towed to their destinations whilst others are 'self-propelling'. The self-propelling type is more prevalent amongst recently built rigs, which are also more likely than older builds to have thrusters for **dynamic positioning**. During propulsion to the scene of use a production rig such as an **FPSO** will not contain a payload, and only the vessel has to be moved. This will be at speeds of the order of 10 m s^{-1}. When it is in service with dynamic positioning it has to resist wave motion at speeds not likely (except under adverse conditions) to exceed 1 m s^{-1} but will at this stage have an actual weight well above the vessel weight. Approximate cancellation of the weight and speed factors means that one would intuitively expect the main engines and those for dynamic positioning to have h.p. values of about the same order of magnitude. As one example amongst many possible ones we consider the FPSO Hanne Knutsen, currently in the North Sea having also seen service in the Baltic Sea. Its 'main engines' have a combined rating of 20MW (26 820 h.p.) and the engines for its dynamic positioning thrusters have a combined rating of 16 MW (21 456 h.p.). Such information is available from web sources for numerous such facilities. A sufficiently interested reader can easily examine data for other FPSOs and the like to confirm or otherwise the author's assertion that cancellation of speed and weight factors in going from propulsion to dynamic positioning leads to comparable power requirements for the two. At Hanne Knutsen, diesel fuel is used for dynamic positioning and (as expected) bunker fuel for propulsion. (*See* **Kovytka Field**.)

Proppant

(*See* **Hydraulic fracturing fluid**.)

PTT PCL

Formerly the Petroleum Authority of Thailand. It is a participant in an undertaking to produce LNG at an **FNGU** off north-western Australia. One FNGU of specification by **FLEX LNG** is currently being constructed in Korea for use in this venture. The LNG it produces will be shipped to Thailand for use there. (*See* **Arthit Field**.)

Pump jack

A.k.a. a beam pump, a sucker rod pump or a nodding donkey: a positive displacement pump used at oil wells as a means of artificial lift. Having been used since the early days of oil production and become over that time an icon, it remains a widely used piece of plant at onshore oil wells. As a single example amongst many, the Lost Hills oil field in California contains several hundred pump jacks and these are one of the 'sights' along State

Highway 46. Similarly in other parts of the US nodding donkeys are part of the landscape. A key quantity in the performance of a pump jack is the torque. The major manufacturer Drake of Sheffield PA, whose products are primarily for the use at the Appalachian oil fields, offers pump jacks having a range of 2000 inch-pounds to 25 000 inch-pounds (225 N m to 2825 N m) torque rating. At an oil well the pump jack will be installed at a wellhead and up-and-down movement of the sucker rod provides the pumping function. In the Drake range referred to, the stroke length is up to 36 inches for the top-of-the-range model. This of course represents penetration into the well, therefore a pump jack is partly an above ground device and partly a below ground device. On descent of the sucker rod pressure in the space rises, and oil is forced out of the formation, entering the space via perforations in the casing and in the cement which holds it in place. At this stage a previously open check valve closes and the rod in its upward movement draws the released oil to the wellhead. Where there is associated gas (as there often will be) this exits via the annulus between the pump barrel and the well casing and is collected at the wellhead. This provides for easy delivery of the gas and prevents interference by the gas in the functioning of the pump jack. Related calculations drawn using only ball park values for the input quantities follow in the box.

Using a value of 36 inches for the stroke and a bore diameter of 1 inch, the swept volume is:

$$\pi \times 0.5^2 \times 36 = 28 \text{ cubic inches or } 0.46 \text{ litres, } 0.0029 \text{ barrels}$$

If the pump jack operates at 10 strokes per minute the oil yield is 0.029 barrels per minute or 42 barrels per day.

The Lost Hills oil field previously referred to produced in 2006 approximately 12 million barrels or:

$(12 \times 10^6/365)$ barrels per day = 33 000 barrels per day requiring 33 000/42 pump jacks = 785 pump jacks, broadly consistent with the figure of 'several hundred' given above.

The calculation can be taken a little further in the following way. Work done in raising 33 000 barrels of oil through a distance of 36 inches =

$$33\,000 \text{ bbl} \times 0.159 \text{ m}^3\text{bbl}^{-1} \times 950 \text{ kg m}^{-3} \times (36 \times 0.0254) \text{ m} \times 9.81 \text{ m s}^{-2}$$

$$= 45 \text{ MJ}$$

where a figure of 950 kg m^{-3} has been used for the density of the crude oil.

$$\text{Rate of work} = 45 \times 10^3 \text{ kJ}/(24 \times 3600) \text{ s} = 0.52 \text{ kW} = 0.69 \text{ h.p.}$$

This is the work developed by all the pump jacks in raising oil by a distance of the pump stroke only and represents about 0.1 of 1% of the typical horse power rating per pump. Conversely, the height by which the oil is raised will be about 900 m according to this rough performance calculation as the interested reader can easily verify. (*See also* **Nitrate ions, action on by SRB**; **Perdido platform**.)

Pumping station

The approximate counterpart for an oil pipeline of a **compressor station** for a gas pipeline. Crude is of course incompressible so a pump is required to achieve the effect that a compressor station has on gas flow. A related calculation follows.

> For a particular oil pipeline in LA, USA, four 6000 h.p. pumps serve a pumping station with a throughput of 2.4 million barrels of oil per day. Now 24 000 h.p. is equivalent to 18 000 kW. The oil passage in thermal terms is:
>
> 2.4×10^6 bbl $\times$ 0.159 m^3 bbl^{-1} $\times$ 950 kg m^{-3} $\times$ 40 $\times$ 10^6 J kg^{-1}/(24 $\times$ 3600) s W
>
> $= 1.7 \times 10^{11}$ W
>
> pump power/hypothetical heat release rate on burning the oil at the pumping rate =
>
> $1.8 \times 10^7/(1.7 \times 10^{11}) \times 100\% = 0.01\%$, one ten-thousandth.

It has been shown that for a compressor station this ratio is about one thousandth, so as far as it is valid to argue on the basis of this single calculation there is better return on a pumping station than on a compressor station, although neither has a heavy energy consumption. (*See also* **Forties pipeline**; **Kazakhstan-China oil pipeline**; **South-East to North-East pipeline**; **Trans-Alaska pipeline**.)

PWD

Pressure while drilling, a form of **MWD**. A PWD instrument will often incorporate a quartz pressure gauge, which works on the principle that electricity is generated within the quartz to a degree dependent upon the forces it is experiencing. A pioneer in PWD was Sperry Drilling, a division of Halliburton. Its recent promotional literature describes a case study involving PWD at the Campos Basin offshore Brazil a précis of which will be given here. At Campos **NPT** was initially heavy owing to factors including **kicks**, vibrations, low **ROP** and the short life expectancy of drill bits. Introduction of PWD enabled the **WOB** to be monitored to the enhancement of the ROP. Well cuttings build-up was also detected, and its potential effects prevented by **drilling fluid** control. There was one incident in which the PWD itself failed as a result of vibrations. The **DDS™** (itself a Halliburton product) was from that point used concurrently with the PWD device. (*See* **Zhana Makat Field**.)

Qalhat LNG plant

In Oman, comprising a single LNG train with a nameplate capacity 3.3 megatonnes per year and using mixed refrigerant. There are also 80 000 tonnes (0.7 million barrels) per year of condensate. There is a contract with the Korean Gas Corporation under the terms of which shipments of LNG from Qalhat are taken to the **Incheon LNG terminal**.

QT-800®

Material for **coiled tubing drilling**, from National Oilwell Varco. It is a steel containing amongst its minor elements copper and on this basis is distinguished from a 'carbon steel'. The importance of the copper content is resistance to corrosion. There are a number of materials in the QT range which have been used in coiled tubing drilling, including QT-700®. One way in which QT-800® is an advance over QT-700® is reduced propensity to 'ovality growth', that is, departure of the cross section from being circular.

Quantec™

Range of drill bits from Baker Hughes, using **PDC** for the cutting surfaces. In recent North Sea drilling in a sandstone formation, a Quantec drilling bit of 8½ inches diameter enabled an **ROP** of almost 12 metres per hour to be realised with very significant financial benefits.

By the same author

Hydrocarbon Process Safety

- '... is very well written in a commendably easy style which should ensure wide and thorough readership. ...I am confident the book could become a standard text for many courses in process safety at both undergraduate and postgraduate level'. ***Professor John W. Patrick, University of Nottingham, UK***
- '... can be used as a text for an introductory course on process safety. It also contains topics relevant to process hazards in the chemical process industries and, as such, will be useful to engineers working in these industries'. ***Journal of Loss Prevention in the Process Industries***

Rabi Field

Oil field in Gabon to which 10% of that country's GDP is due. There is associated gas, and consultants were brought in when that began to be a problem because of reservoir pressure decline. Gas reinjection, not of course a new technology, was introduced with an increase in oil production from 55 000 barrels to 65 000 barrels per day. This is still a long way below the 225 000 barrels per day which the field was once producing and it is expected that a means of artificial lift will sooner or later be introduced. (*See also* **Etame Field**; **Olowi Field**.)

Rajasthan block

In north-west India, producing oil since 2009 with continuing development. The operator is Cairn India. Many wells have been drilled, to depths of typically 1630 m, not all of which have led to oil in viable amounts. The formation varies quite widely in permeability. The most productive of the wells so far drilled is called Mangala, and can produce 10 000 barrels of crude oil per day. Many of the procedures for oil field development are taking place there including **directional drilling** with **LWD** and **MWD** and hydraulic fracture. Power available for the hydraulic fracture is 8000 h.p. **Electric submersible pumps** are in use at some wells and, at least one, downhole heating. An offshore drilling rig is either moored or dynamically positioned, drill access being via the **moonpool**. With onshore drilling the rig supporting the drill might have to move to accommodate forces in the drilling assembly. This is called an *intra-pad move*, and it is recorded that over a period of 301 days there were 37 such moves at Rajasthan. An intra-pad move is of course made possible by the wheels supporting the rig and is a time-consuming operation. It was made less so at Rajasthan by the fact that such movement was possible without lowering the mast of the drilling rig ('rigging down'). (*See also* **Zuata Field**; **Ravva Field**.)

Ram-type blowout preventer

This uses a plate equivalent in area to the cross section of the well in which it is installed. This plate is, in normal well operation, in two halves that are brought together by hydraulic machinery when the device is activated. Its action is therefore analogous to the closing of a gate valve. It will use elastomer seals where any two metal components of the composite well-BOP structure are in contact.

Raroa[55]

FPSO in New Zealand waters, converted at Jurong Island from an oil tanker. It has been in service off the west coast of the North Island since 2009. It has mooring lines connected to a buoy on the seabed, an arrangement called submerged turret production (**STP**). The turret is on the buoy and therefore 'submerged'. Mooring and riser systems originate at the submerged buoy. Raroa operates in shallow water (about 100 m) and can hold 0.6 million barrels of oil. It is expected to be in service for approximately of 15 years. In November 2010 there was a small oil spill at Raroa.

Rasau Field

In Brunei, having been producing oil since 1983 and now with 27 production wells at depths in the range 1490 m to 1980 m. The formation is sandstone of permeability in the range 1 millidarcy to 100 millidarcy. The other onshore field in Brunei is the Seria Field, which has been producing since 1929. At that time there was a large expatriate British community in that region of the world working in oil. Seria produced its billionth barrel in 1991. Current production there is 10 000 barrels of oil per day. This is taken initially to a terminal and from there to a tanker held in situ for oil transfer by two single buoy moorings. (*See* **Lumut**.)

Rashid Field

Oilfield offshore the UAE, discovered in 1973. In 2008 the Maersk Resilient, a **jack-up**, was in drilling service there when a helicopter attempting take off from its helideck crashed, with the loss of seven lives. (*See* **Al Jalila Field**.)

Raton Basin

Source of **CBM**, mostly in CO but extending over the border to NM. Its wells (459 in number) deliver much more produced water than do those in the **San Juan Basin NM**, which is south-west of Raton and also crosses the NM-CO border. Raton Basin will be used as introduction and background to the matter of produced water in CBM generally. Relevant figures for CBM from wells at the Raton Basin are given below.

Water production/barrels per day per well	266
Water-to-gas ratio/barrels per million cubic feet	1.34 (48 barrels per million cubic metres)
TDS/milligrams per litre	2000

The corresponding figures for the San Juan Basin are:

Water production/barrels per day per well	25
Water-to-gas ratio/barrels per million cubic feet	0.031 (1.1 barrels per million cubic metres)
Total dissolved solids(**TDS**)/milligrams per litre	24000

The following points can be made in respect of TDS. First, even the high value for the San Juan Basin NM is at the low end of the range given in the entry for salinity of produced water from oil wells. Secondly, the ions constituting TDS include Na^+, HCO_3^- and Cl^-. Thirdly, over half the produced water from CBM production in the US is disposed of at specially drilled 'disposal wells'. Such water can be used in hydraulic fracturing operations and this has in fact been done by Halliburton. Produced water from CBM sources is often called 'pumped water', meaning of course that it has been pumped *from* the coal seam. (*See* **USA, selected CBM reserves in**.)

Ravenspurn North

Gas field in the UK sector of the North Sea, operated by BP and the scene of a **GBS** for production. It has a **POB** of 30 and is connected to two unmanned 'satellite platforms'. The primary platform and its 'satellites' collectively are connected to 42 wells and there are enough well slots between the three for that to be increased to 54. (*See* **Dimlington**.)

Ravva Field

Oil and gas field offshore India operated by Cairn India, as is **Rajasthan block**. Believed a few years ago to be depleting, the field has exceeded predictions of recent performance having produced 27.5 thousand barrels a day during May 2010 against a target 21 thousand barrels per day. To this has to be added major amounts of associated gas. There are eight production platforms at the field.

Reaming

The widening of a well bore where the diameter is below that expected, usually through wear to the outer surface of the drill bit. Such wear will obviously be most likely to occur late in the life of a drill bit, and reaming when it is required tends to be close to the **BHA** where the structure is highly sensitive to vibration. This makes reaming close to a BHA a delicate procedure. A reaming device might be two-component: a support ('shoe') to which a suitable drill bit is attached. (*See also* **DiamondbackTM**; **NBR®**; **XRTM**.)

Red Hawk

The first **spar platform** in the Gulf of Mexico, in 1615 m of seawater with polyester fibre leads instead of the more usual steel ones. Linked to anchors on the sea-floor, these withstood Hurricane Katrina. Use of polyesters in 'mooring lines' is increasing, not only with production platforms but with **semi-submersible** rigs involved in drilling, e.g. **Ocean Victory**. The **FPSO** in service at the **Barracuda Field** uses polyester lines as does **Fluminense**. (*See* **Independence Hub**.)

Reel laying of pipelines

Having been the norm in the early days of offshore oil production this went out of use in favour of pipe sections. There has been limited revival, for example **Deep Blue** is equipped for reel laying as well as for **J-Lay** of pipe sections. With reel laying, pipe is carried on the barge as a spool instead of in straight sections. Lengths taken from the spool need welding before installation, although other things being equal less onboard welding is needed with reel laying than when pipe sections are used. Mounting of the pipe onto the spool is done onshore (where labour costs are inevitably lower than offshore) and the spool taken out to the **pipe laying vessel** by a supply vessel, and this has some advantages. Once a section is taken from a spool it is a 'candidate' for **S-Lay** or for J-Lay according to requirements and conditions.

Regrinding of well cuttings

Once a fragment of the formation has been removed by drilling, its prompt removal from the vicinity of the drill bit is desirable. If it remains close to the drill bit for a significant time it will undergo regrinding, which is of no value and represents a waste of power and of cutting material. Recent bit development by Varel has applied computational fluid dynamics (CFD) to the drill fluid in order to identify conditions leading to regrinding and to avoid them. In the decommissioning and eventual 'abandonment' of offshore facilities, well cuttings have to be considered. However, experience in this regard is often encouraging, and when the Welland platform in the North Sea was recently decommissioned it was found that cuttings were nowhere any greater than 2 mm in thickness and could be left in place.

Rerun tricones

A tricone drill bit having experienced sufficient wear to preclude its continued use in oil drilling can be used in the milder conditions of water drilling, thereby extending the life of a resource. A tricone bit having been reclassified for water drilling is termed a rerun tricone and several major manufacturers of drill bits for the oil industry take back tricones for this purpose. A particular tricone will be examined before being accepted as a rerun, and the only reconditioning it receives is a coat of paint.

RHEO-LOGIC™

Synthetic based drilling mud from Baker Hughes. Introduced in mid-2007, it can be supplied to varied specifications of viscosity and density by use of other Baker Hughes products.

Rhum Field

Gas field in the UK sector of the North Sea in a water depth of 110 m and producing since 2005. BP is the operator. The gas is not of the highest quality, containing appreciable H_2S (making it sour) and CO_2 (making it corrosive). It contains significant condensate. Pipes containing gas from the four production wells at Rhum are tied back to infrastructure at the **Bruce Field** which is also operated by BP.

RING-O®

Ball valve for subsea use from Cameron (HQ in Houston). It is expected to operate without maintenance for 40 years, that being an estimate of how long a subsea pipeline will be in use. The valve in its mechanism is similar to those from the same manufacturer for general use, with the addition of reinforcements to enable the valve to withstand pressures due to the sea depth.

Rio de Janeiro-Belo Horizonte Gas Pipeline II

A.k.a. Gasbel II, entering service in 2010, a pipeline entirely within Brazil. It is 267 km long and has three **compressor stations**. The pipeline diameter is 18 inches and delivery is 5 million cubic metres per day. These figures are examined in the calculations below.

$$\text{Speed of flow of the gas} = \{[5 \times 10^6/(24 \times 3600)]\ \text{m}^3\ \text{s}^{-1}/[\pi \times (9 \times 0.0254)^2\ \text{m}^2\}\ \text{m s}^{-1} = 350\ \text{m s}^{-1}$$

Using a value of 1.4×10^{-5} m^2 s^{-1} for the kinematic viscosity of methane gives a Reynolds number of:

$$(350 \times 18 \times 0.0254/1.4 \times 10^{-5}) = 10^7$$

and the reader can confirm from web sources that Reynolds numbers this high *do* characterize natural gas flow in pipelines.

In examining the above we might observe that three compressor stations along a pipeline 267 km (165 miles approx) is unusual and that this is reflected in the high speed. Whilst as noted the Reynolds number is of a reasonable magnitude, values in the range 5×10^5 to 10^6 are probably more common. Performing a calculation like those for **South-East to North-East pipeline** and **Texas Eastern Transmission** we should expect the combined rating of the compressor stations at Rio de Janeiro-Belo Horizonte to be around 2000 kW, a little under 2700 h.p. The gas from the pipeline goes to the Brazilian state of Minas Gerais, which also receives gas from other pipelines including Gasbel I (capacity 3.1 million cubic metres per day). Populous (19.5 million) and industrialised (for example, Fiat and Mercedes Benz each have a plant there), Minas Gerais has a total natural gas requirement of 12.9 million cubic metres per day.

Rio Vijo Field

Oil field in California, one well at which - Batson 87X-34 - has a depth of 4.4 km making it the deepest currently producing well in California. The **pay** at the well is a mere 21 m.

Rivers Fields

Collective name for five fields - Calder, Darwen, Crossans, Hodder and Asland - in the eastern Irish Sea. Calder is the largest and has to date seen the most development. One well at Calder is vertical with depth 914 m. The other two were drilled directionally and have measured depths (MD) in excess of 2000 m. There is a **fixed platform** at Calder to which wells at the other fields in the group will be tied back. A 50 km, 24 inch diameter, pipeline takes the gas to an onshore treatment plant near Barrow-in-Furness. Such 'treatment' includes hydrogen sulphide removal, for the gas is very sour. Beyond there the gas uses the infrastructure at the Morcambe Bay Field.

RMU

Refrigerant Manufacturing Unit. Light hydrocarbon compounds do of course find application to refrigeration cycles, their vapour pressures and normal boiling points being suitable for such use. When at an LNG plant such compounds are stripped off and separated, they can be used on site as refrigerants in which case the stripping/condensation facility is called an RMU. It has been explained how use of such refrigerants eliminates the need to get methane below its critical temperature by Joule-Thomson cooling. (*See* **Lumut**.)

Rocknes

DPFV. It sank off the Norwegian coast in January 2004. Eighteen of its crew of thirty lost their lives. It was carrying a payload of rock at the time.

Roller cone drill bit

Invented by Howard Hughes himself, this device has cones on which spiked teeth are mounted. These cones undergo movement additional to that of the drill structure ('drill string') in contrast to a *fixed cutter bit* which simply rotates at the speed of the drill stem. (*See also* **Triton™**; **Tuffcutter™**.)

Roller reamer

Downhole tool, a.k.a. a rotary reamer. It has two functions: **reaming** the well and, by reason of its contact with the bore, stabilising the drill string. A roller reamer will have a body, incorporated with the drill string, with several cutters which are supported by bearings enabling them to 'roll'. The cutters are most commonly steel, although with hard formations a **tungsten carbide** insert might be used. The cutters, arranged symmetrically around the body, can be replaced as required eliminating the need to obtain a whole new tool. In some designs the cutting surfaces are arranged spirally. A roller reamer *can* be part of a near-bit reaming arrangement but will not necessarily be.

Rollingstone

Fall pipe vessel operated by Tideway and registered in the Netherlands. It has **dynamic positioning**, and admits rock to the sea through a lined 8 metre 'fall pipe' at a **moonpool**. It was built in 1979 and adapted to its current function in 1992. (*See* **Malampaya Field**.)

Roma to Brisbane pipeline

This conveys gas from Roma, in the Darling Downs region of Queensland, Australia, to the state capital Brisbane over a 434 km length. The piping sections were made by the same firm (themselves headquartered in Roma) as made those for the **Bonaparte gas pipeline**. One looped part of the Roma to Brisbane pipeline of 13 km length is made of **X80 steel**.

Rong Doi Field

Condensate field offshore Vietnam. It produces 4.6 million cubic metres of gas and 6000 barrels of condensate per day. The gas is taken ashore by the **Nam Con Son pipeline**.

ROP

Rate of penetration in well drilling. It is probably impossible to give a 'typical value' but tens of feet per hour would be the order of magnitude commonly experienced. A value below 30 feet per hour might be seen as inconveniently low. Accordingly there are 'ROP enhancers', one such being the Halliburton product XLR-RATE™. This is made from synthetic fluids and, like other ROP enhancers, works largely by reducing friction at the drill bit. The data sheet for XLR-RATE™ reports a rise in ROP in one particular subsea application from 25 feet per hour without an enhancer to 41.2 feet per hour with 3.0% XLR-RATE™. XLR-RATE™ is particularly suitable for use with **PDC** drill bits. In general the ROP depends on the hardness of rock and in the North Sea chalk formations have been known to make for a low ROP. Recent use of a drill bit from the **Quantec™** range effected an improvement in this regard. (*See also* **Drill bit, life expectancy of ('bit life')**; **F21 Well (BP)**; **Laos, oil exploration in**; **Quantec™**; **Shearwater Field**; **Shock absorber**; **Spear drill bit**; **Tiber Field**; **Tuffcutter™**; **Woodford OK**.)

Rosa Field

Oil field offshore Angola, water depth 1350 m. It was discovered in 1998 and began producing in 2007. Oil from the field goes to the **Girrasol** facility, to which 25 wells from 15 km away – 14 for production and 11 for **water injection** – are tied back. (*See* **Farwah**.)

Rosneft, selected major upstream assets of

Rosneft, HQ in Moscow, produced in 2009 almost 800 million barrels of oil as well as natural gas and condensate. Details of a selection of oil and/or gas fields in which Rosneft or one of its many subsidiaries have involvement are given in the table below. Comments follow the table.

Field name and location	Details
Priobskoye Field, Western Siberia.	Occupying an area of 5466 square kilometres. Production commenced in 2000, and there has recently been hydraulic fracture at the field. The field is operated by Yuganskneftegaz which, previously independent, was acquired by Rosneft in 2005.
Sovetskoye Field, Western Siberia.	Producing since 1962 and now requiring redevelopment. It is operated by Tomskneft, which was acquired by Rosneft in 2007.

Vankor Field, Eastern Siberia.	Entering production in 2009, and one of the major 'finds' of the previous 20 years. Production at commencement 130 000 barrels per day. A peak production of over half a million barrels per day aimed for as more wells are drilled.
Vereiskian Field, central Russia.	Multiple reservoirs and a **gas cap**. Recent development by sidetracking of wells.
West Ozyornoye gas field, Southern Russia.	Operated by Sibneft, in which Rosneft and Gazprom each have an interest. Production 120 million cubic metres of gas per year from four production wells.
Yuzhno Russkoye gas field, Western Siberia.	Producing since 2009 from 142 production wells each of depth about 1000 m. The gas is of good composition being 98% methane, and is exported to Germany.

Some of the hydraulic fracture work at Priobskoye was contracted to Halliburton and some to Newco, the Russian subsidiary of Trican Well Services (HQ in Canada). Halliburton have a base in Poykovskiy, about 60 miles from the scene of their operations at Priobskoye. It is reported that in the Halliburton operation 600 000 pounds (272 tonnes) of **proppant** were used. The redevelopment at Sovetskoye (row 2) will include well workover and sidetracking of some wells to extend their range. Oil from Vankor is taken to the ESPO (East Siberia Pacific Ocean) pipeline to a terminal at Kozmino whence it is shipped to countries including South Korea and Japan. Sinopec, in addition to Rosneft, have had an interest in Vereiskian. Some of the sidetracks referred to at Vereiskian extended half a kilometre from the original well before encountering productive formation. The sidetracking raised production at the field to about 120 000 barrels per day. Whether some of the gas from West Ozyornoye could viably be converted to LNG is being considered. Meanwhile some of it goes along a 104 km pipeline to a power station. Rosneft, with many other companies, is active at Sakhalin in activities including the Odoptu project. (*See also* **Caspian pipeline consortium**; **Kovytka Field**; **Samotlor Field**; **Severo-Varyoganskoe inter alia**.)

Rowan Gorilla III

Jack-up drilling rig having recently been in use at **Deep Panuke Field** having been towed there. It can drill in water depths up to about 135 m and to well depths of about 9000 m. Its **derrick** can take a load of 565 tonnes and its **drawworks** are rated at 3000 h.p. It has often been used in the Gulf of Mexico, and carries the US flag.

RP™ II

Drag reducing agent for pipelines carrying refined products ('RP'), from ConocoPhillips. It is polymeric and is used in amounts up to 60 p.p.m. There have been comprehensive tests using both Ford and GM engines which demonstrate that use of the additive at up to that level in a pipeline conveying gasoline has no effect on subsequent automotive use of the gasoline.

Rubicon Offshore, vessels operated by

Rubicon Offshore, based in Singapore, was formed in 2005 and many of its vessels are rebuilds, predating the company in their pre-rebuild service. The vessels operated by Rubicon are described in the table below.

Vessel name	Description
Rubicon Intrepid	FPSO, rebuild of a vessel having first entered service 1981, supplied to Rubicon in 2007. **POB** 32. Oil production capability 24 000 barrels per day. Oil storage capacity 0.45 million barrels. Single point mooring with DP for station keeping. A burner for gas flaring.
Rubicon Vantage	FPSO, have been used at the **Bualuang Field** as noted. Rebuild (2007). Oil production capacity 44000 barrels per day. Oil storage capacity 0.55 million barrels. **Water injection** capability at 55000 barrels per day.
Front Puffin	FPSO, entering service as a rebuild with Rubicon in 2007. Oil production capacity 40 000 barrels of oil per day. Oil storage capacity 0.73 million barrels. Submerged turret (**STP**).
Crystal Ocean	FPSO, a rebuild. POB 32. Oil production capacity 37 700 barrels per day. Oil storage capacity only 42 000 barrels so if operation is at maximum production transfer to a tanker or a storage facility will need to be very prompt.
Sea Cat	'FPSO candidate', that is, currently undergoing conversion. Some features 'to be decided'.
Rubicon Maverick	Multi-purpose Support Vessel (MSV), built in Germany in 2001 and acquired by Rubicon in 2007 and modified to their specifications, for example to incorporate a **moonpool** so that well intervention duties can be undertaken. POB 88.

The table above summarises the entire fleet operated by Rubicon, no member of which was initially built expressly for the company. Rubicon Intrepid has recently been in service at the **Galoc Field** together with Rubicon Maverick. Crystal Ocean and Front Puffin are both at the time of going to press working in Australian waters.

Rudong LNG terminal

In the Jiangsu province of China, operated by Petrochina it received its first shipment of LNG in May 2011. This was from Qatar, and there is a 25 year pact whereby the terminal will receive from Qatar. It can receive 3.5 million tonnes annually. By a calculation like that given in detail for the **Dragon LNG Terminal** it can be shown that this corresponds to a gas supply of 13 million cubic metres per day. (*See also* **Fos Tonkin (Fos-Sur-Mer) LNG Terminal**; **Myanmar-China pipeline project**; **Pioneer Knutsen**.)

Rudong LNG terminal
(courtesy: Simon Q)

RwC™

Reaming with casing, a process developed by Weatherford whereby instead of drilling followed by casing, a section of casing is attached to a reamer shoe and therefore (obviously) rotates with it. With a suitable cutting surface most likely to contain **PDC** the composite structure penetrates a formation leaving the casing in place. It is normally only used where the more regular procedure is made difficult by formation instability. A recent case study appertains to an onshore field in Culberson County TX. A horizontal distance of 302 m was penetrated by application of RwC™. The **Diamondback™** reamer shoe has proved its worth in RwC™, notably in an operation in Papua New Guinea where the run length was 355 m and took in a **fault**.

S

Sabine Pass

Location of the **Golden Pass LNG terminal** and of the Sabine Pass LNG terminal, operated respectively by ExxonMobil and Cheniere. An attraction of this part of the Gulf coast for LNG terminals is the water depth facilitating tanker entry. Gas once obtained as such from LNG at Sabine is marketed and principal purchasers are Chevron and Total. Sabine Pass is one of only two LNG terminals with the authority to re-export LNG once received, and three shipments have recently been so exported from Sabine pass, one each to the UK, Spain and India. The other US LNG terminal with the authority to re-export is that at Freeport TX which is 50% owned by ConocoPhillips.

Sacha 171H well

Recently drilled production well at the **Shushufindi Field** in Ecuador. It was drilled horizontally so that infrastructure already in place was accessible to oil having exited the well. It is producing 6000 barrels per day. It is one of four recently introduced production wells at the Shushufindi Field, all of which were obtained by horizontal drilling. The others are Sacha 175H (2260 barrels per day), Sacha 173 (948 barrels per day) and 102H (1454 barrels per day).

SAFE-CARB

Family of bridging agents manufactured by a company in the Schlumberger group, made of calcium carbonate. A suffix numeral denotes the particle size range. Obviously such a substance influences the mud density, and its roles as a densifying agent and as a bridging agent have to be held in balance when the quantity added to a drilling mud, to a drill-in fluid or to a completion fluid is determined. A sister product for the same purpose is TRUCARB. In this the calcium carbonate particles have been coated to make them organophilic.

SAFE-VIS™

Viscosifier for **completion** fluids, a Schlumberger product. It is a polymeric substance which becomes dispersed in the brine base of the gel rather than suspended in it, as, for example, **bentonite** would be, making it resistant to deposition and therefore suitable for use under the very fragile conditions of reservoir entry. Even so, filtering of a completion fluid to which SAFE-VIS™ has been added might be necessary to remove globules ('fish eyes') formed through the incomplete dispersion which results when not all of the added product becomes hydrated.

St John, New Brunswick

Scene of an LNG terminal in service since 2009. It is capable of providing from LNG 28 million cubic metres of natural gas per day, a 40% margin over the **Fos Tonkin (Fos-Sur-Mer) LNG Terminal**. It receives LNG from Trinidad and Tobago. Its first shipment, fully recorded photographically on the web pages of the operator Canaport, was from the **Bilbao Knutsen, Kiskadee Field**.

Saipem 7000

Pipe laying vessel. There was a fatal accident on board the vessel in 2008 when it was working in the Mediterranean, laying pipe in a **J-Lay** fashion.

Saipem 10000

Drill ship, built in Korea by Samsung and entering service in 2000. The numeral in its name is the nominal water depth in feet at which it can drill and is equivalent to 3050 m making Saipem an 'ultra deep water' facility. It can drill to 6000 m **BKB** and has **dynamic positioning**. Its **drawworks** are rated at 4500 h.p. Its **derrick** has a height of 200 feet and can support a load of 900 tonnes. It has seen recent use offshore Angola where drilling was for the Italian oil company Eni.

Sakhalin, LNG production at

LNG production at Sakhalin began in early 2009 and is at the Prigorodnoye production Complex at Sakhalin. In the liquefaction step the Double Mixed Refrigerant method, developed by Shell, is applied. Initial cooling is with a *mixed refrigerant*, actually ethane and propane, and the spiral wound heat exchanger (SWHE) is used in both initial cooling and liquefaction steps in contrast to practice at the **Snøhvit LNG facility** where the SWHE is used only in liquefaction. This entry of course refers to gas converted to LNG *at* Sakhalin. Much gas leaving Sakhalin as such is destined for conversion to LNG further downstream. (*See also* **Liquefied natural gas (LNG), manufacture of**; **Qalhat LNG plant**; **Rosneft, selected major upstream assets of**.)

Salah Gas

Joint activity between BP and Sonatrach (Algeria) in which tight gas reserves in Algeria are being evaluated. Any production will add to the already very major (1.4 billion cubic metres per year) conventional natural gas production in Algeria. Hydraulic fracture is made more effective if it is directed at a natural fracture, in which case the term 'stimulated natural fracture' can be applied. Part of the appraisal work for Salah tight gas is the identification of natural fractures in the formation. (*See also* **Shale natural gas**; **Toot Field**.)

Salinity, of produced water

For oil reservoirs in the US this is in the very wide range 100 mg L^{-1} to 400000 mg L^{-1} where the milligram refers to mass of sodium chloride. A simple calculation follows.

The geometric mean of the limiting values is $(100 \times 400\ 000)^{0.5}$ mg L^{-1} = 6325 mg L^{-1}.

This is a factor of five to six below the salinity of seawater.

In the entry for **Icebreaker Sakhalin** it was stated that sea temperatures encountered could be as low as −40 °C, representing a freezing point depression of 40 °C. Using a value of 1.86 °C (mol $L^{-1})^{-1}$ for the freezing point depression constant of water this gives a concentration of depressant of:

$$40/1.86 = 22 \text{ mol L}^{-1}$$

Now this is moles per litre of ions, Na^+ and Cl^- equally. The concentration of sodium chloride is therefore half of this or 11 mol L^{-1}, equivalent to 6.4×10^5 milligrams per litre. This is comparable to the upper limit for produced water and two orders of magnitude higher than the geometric mean value. It is four and a half orders of magnitude higher than the lower limit for produced water.

One would not regard water from an aquifer, which is not saline at all, as produced water. Where in well drilling there is an aquifer it will be seen as an asset in providing downhole fresh water and protection from contamination by drilling mud and the like. It is described in the entry for **Buckland Field** how judicious use was made of water from an aquifer there. Produced water can be used instead of seawater in **water injection**. If the produced water is less saline than seawater the effect on the formation of the injection is less because there are fewer minerals and inorganics for subsequent deposition. At the Ghawar Field in Saudi Arabia there is significant reinjection of produced water and, consistently with what has been said in the previous sentence, it is seen as being of greatest protective value in the parts of the field with lower porosities. (*See also* **Aral Sea**; **Bolivar Coastal Field**; **Raton Basin**.

Salt Flat Field TX

Productive from 1928 to the 1960s when it became uneconomic to operate. The formation is limestone. In 2007, by which time horizontal drilling had become well established, the field was acquired by developers with a view to its re-entering production. Oil at the field is about 850 m below the surface and in two **pay** zones, one of 15 m and the other of 45 m. Porosities of the formation are in the range 10% to 35% and it has been confirmed that the formation is undamaged in spite of its long previous period of production and will not therefore require stimulation by acid. All of this augurs well for the prospects of Salt Flat Field.

Samotlor Field

In Siberia, the largest oil field in Russia. It was discovered in 1965. Reservoir depths are 1500 m to 2000 m. Having by the approach of the 21st Century been deemed uneconomic to remain in production, the field is expected to undergo restoration which will involve both hydraulic fracture and *multilateral wells*, that is, wells with multiple sidetracks. Horizontal wells are also being considered. It is operated by TNK-BP. (*See* **Pickerill Field**.)

San Juan Basin NM

The scene of the production of **CBM**. It has been chosen for coverage as an example of how products and techniques developed for oil production have also been used in CBM production. Halliburton's **Sandwedge**® conductivity enhancement method was used at ten wells there as was DeltaFrac, **a hydraulic fracturing fluid** from the same company. Perforation at **completion** was four shots per foot. CBM yield from the wells at the San Juan Basin having received such treatment is 0.41 million cubic metres per day. (*See also* **Huabei Field**; **Raton Basin**; **USA, selected CBM reserves in**.)

SandWedge®

Process by Halliburton for enhancing the conductivity of a **proppant**. Its action is directed at ingress of particles from the geological formation into the proppant which over time causes loss of conductivity. Such ingress is influenced by the bulk density of proppant within a fracture which in turn is influenced by the settlement rate of particles. SandWedge® acts by decelerating settlement. It also inhibits diagenesis, that is, chemical reaction between proppant and formation which, at the temperature fracture site, can be significant and have a strong negative effect on conductivity. (*See also* **San Jaun Basin NM**.)

Sanga-Sanga block

In East Kalimantan, a source of **CBM** which, if realised, will go to the **Bontang LNG plant** for liquefaction. As at **Curtis Island** (or anywhere where CBM becomes LNG) there will be no supplementary products from natural gas liquids as there are when conventional gas is made into LNG. Participating companies at Sanga-Sanga include Eni and BP. (*See also* **Barito Basin**; **Jatibarang Field**.)

Sangaw North

Exploration well in the part of Iraq currently seen as being Kurdistan territory. **Spudding in** took place in early 2010 and the target depth is 3660 m. The operator is Sterling Energy. Also in Kurdistan is the Taq Taq Field, being explored by Addax Petroleum. The drilling rig Kurdistan-1 is operating there at the present time. Also in 'Iraqi Kurdistan' is the *Tawke Field*, now producing at 20 000 barrels per day. The oil is set for export to the west. This has been delayed by a decree from the Iraqi government on export from fields currently in Kurdish possession. There is not yet a permanent territorial Kurdistan.

Sapele well

A.k.a. the Sapele 1 Exploration well, in the Douala Basin[56] offshore Cameroon in a water depth of 25 m. Directionally drilled, the well reached a **TVD** of 3634 m and a measured depth of 4483 m, giving an **AHD** of:

$$[4483^2 - 3634^2]^{0.5}\text{ m} = 2625\text{ m}$$

The **jack-up** named **Noble Tommy Craighead** was used in the drilling and the pay of hydrocarbon is estimated to be in the range 30 m to 40 m. A sidetrack ('Sapele 1ST') from the well has revealed further **pay** of about 23 m. The operator is Bowleven, HQ in Scotland, whose web site at the time of writing reports that a second exploration well (Sapele 2) is 'progressing on schedule'. (*See* **Logbaba**.)

Sarah

Well intervention vessel completed in 2009. It and its 'twin sister' Karianne differ from most other such vessels in having an **X-Bow** rather than a semi-submersible structure and the following specifications for Sarah also apply to Karianne. The **POB** is 100 and there is **dynamic positioning**. It can 'intervene' at well depths up to 2500 m. One expects an intervention vessel,

Marine Subsea's vessel, Sarah
(courtesy: Helen Hadley)

the purpose of which is to provide tool access to subsea structures, to be involved in decommissioning. Sarah arrived in Aberdeen in August 2010 in order to undertake some decommissioning contracts in the North Sea. (*See* **Tristan North West**.)

SARLAM

Single arm rigid leg anchor mooring, a form of single point mooring in which part of the mooring line is rigid, being neither a metal rope nor a chain. (*See* **Challis Venture**.)

Scarab and Saffron Fields

Offshore Egypt, producing gas since 2003 and undergoing development. The water depth varies from 250 m to 850 m. It is 90 km from *terra firma* at the Nile Delta where there will be facilities for receiving up to about 20 million cubic metres per day of gas and for **condensate stripping**. Pipelines for the gas were laid by **LB 200**, a **pipe laying vessel** operated by the former Stolt Offshore, now Acergy.

Schiehallion

Oil field in the western North Sea. Discovery of the field was with the **semi-submersible rig** Ocean Alliance. The water depth is 400 m and the operators include BP and Amerada Hess. A specially built and commissioned **FPSO** that can store 0.95 million barrels of crude oil and has **water injection** capability is present at Schiehallion.

SCV

Submerged Combustion Vaporiser used with LNG as well as with other cryogens. The post-combustion gas from burning natural gas is used to heat water containing a metal spiral through which the LNG is passed, with the effect of 'regasifying' it. The acronym SMR – submerged vaporiser – is equivalent. (*See also* **Bahia de Bizkaia Regasification Plant**; **Grain LNG Terminal**; **Hazira LNG terminal**; **Incheon LNG terminal**; **Maasvlakte LNG terminal**; **Mizushima LNG import terminal**; **Sodegaura LNG terminal**.)

Seaharvester

A 'minimum facilities' platform developed in the US, several of which have since been made in the UK under licence for North Sea use. Those for North Sea use have better protection against ship impact than those built in the US for use in the Gulf of Mexico. The UK manufacturer states that either of the examples at the **Boulton Field** could take 'an impact from a 3700 tonne supply ship arriving at about 2 m s^{-1}'. In this event the kinetic energy at impact would be:

$$(0.5 \times 3700 \text{ t} \times 1000 \text{ kg t}^{-1} \times 4 \text{ m}^2 \text{ s}^{-2}) \text{ J} = 7.5 \text{ MJ}$$

which is equivalent to the blast energy released by about 2 kg of TNT.

Seahorse

Fall pipe vessel, said to be the largest such in the world[57] and operated by Boskalis. Entering service in 2000 and having **dynamic positioning**, it admits rock to the sea via an 8 metre diameter lined pipe at the **moonpool**, which is immersed close to the scene of the intended location of the rock. Also operated by Boskalis is Sandpiper, a **fall pipe vessel** considerably older than Seahorse and a rebuild. Like Seahorse, Sand Piper is a **DPFV** and the latter uses a fall pipe composed of polyethylene segments as does **Nordnes**. Both vessels were deployed in the development of satellite fields at **Gullfaks**, where they were involved in laying gravel support for pipelines.

Secondary cutters

Additional in a drill bit to a primary cutter, these are placed behind the latter. If at one site on the bit the primary cutter fails the secondary one will in some degree do its job. More importantly it will contribute the part of the **WOB** previously provided by the failed part. In the absence of the secondary cutter this would have added to the load on the other parts of the bit quite possibly leading to breakage. (*See* **FX™**.)

Sedneth 701

Semi-submersible rig, part of the Transocean fleet. It was built in 1973, upgraded in 1994 and received SPS Overlay repairs in August 2009. It can operate in water depths up to 457 m and can drill to well depths of 7620 m. Its **moonpool** is 35 feet by 61 feet. Its derrick loading is 545 tonnes and its **drawworks** are rated at 3000 h.p. It is moored by eight chain lengths each attached to an anchor of weight 12 tons. Sedneth 701 is currently in long-term use in Namibian waters; the 2009 repairs referred to were in readiness for this. (*See also* **Conkouati**; **Transocean Arctic**.)

Sekayu

In South Sumatra, the scene of a **CBM** well having recently undergone **spudding in**. The target depth for **completion** is 610 m. It is the first 'spudding' of a CBM well in the whole of Indonesia, which is surprising as the country has so much coal and is indeed a major exporter. A published (in September 2009) estimate of the CBM reserves of Indonesia is 12 cubic terrametres (Tm^3). (*See* **Sanga-Sanga block**.)

Semi-submersible rig

Such a rig contains cylinders called 'pontoons', possibly in a catamaran structure (as with Deepwater Horizon, where the 2010 Gulf Coast oil spill happened). These double up as a means of moving the rig when it is not in operation; when the pontoons contain air only they enable it to float and be propelled or towed. For drilling, seawater is admitted to the pontoons and this causes 'submersion'. Having been submerged to the degree required the rig can be held in place by anchors on the sea-floor put in position by an anchor handling (AH) vessel. An alternative to anchorage, and common in deeper water applications, is **dynamic positioning**. The semi-submersible rig has become the most prevalent type of drilling rig, and the semi-submersible design is used not only in drilling but also, for example, in crane support. It can be used in production on a shorter term basis or even for an indefinite term as at the **Gjøa Field**. A semi-submersible rig in such use is often simply referred to as a 'floating production platform' as at the **Njord Field**. In a sense the semi-submersible rig is an intermediate between two other types of rig: the submersible, which in operation touches the sea-floor and the *drill ship* such as **Chikyu** in which there is no submersion at all beyond the natural one. (*See also* **Appaloosa Field**; **Atwood Hunter**; **Big Foot Field**; **Blackford Dolphin**; **Borgland Dolphin**; **Borgsten Dolphin**; **Byford Dolphin**; **Cajun Express**; **Droshky Field**; **DSS-20-CAS-M**; **GSF Grand Banks**; **GSP Bigfoot 1**; **Independence Hub**; **Janice Field**; **Kaskida Field**; **Khazar**; **Nanhai II**; **Petrobras XXI**; **Piranema Field**; **Sedneth 701**; **Shelley Field**;

Shtokman gas condensate; **Stena Spey**; **Stena Don**; **Tahoe**; **Toroa well**; **Transocean Arctic**; **Transocean John Shaw**; **Troll Field**; **Veslefrikk Field**; **Visund Field**; **Well Enhancer**; **WilPhoenix**.)

Sepiolite

Mineral compound of formula:

$$Mg_4Si_6O_{15}(OH)_2 \cdot 6H_2O$$

used in drilling muds. It is capable of a **clay yield** of 20 m to 25 m^3 t^{-1}. It is used in seawater muds and is resistant to coagulation.

Sevan Piranema

FPSO built in 2006–2007 having seen its entire service to date in the **Piranema Field** at a water depth in excess of 1000 m. **Sevan Piranema** is unique amongst FPSOs in that its hull is cylindrical. It is conventionally moored with chain and polyester. It can produce at up to 30 000 barrels per day and can store up to 250 000 barrels. Those are the performance specifications: in the first 2¼ years it produced 6 million barrels, making the actual production rate only 7300 barrels per day with associated gas for reinjection. There have been only 30 visits from **shuttle tankers** over the 2¼ year period. This suggests that the Piranema Field is at present well over-capitalised with production capability, but as noted the field is expanding. There are currently only three production wells delivering oil to Sevan Piranema but there is scope for tying back to it new wells drilled by Ocean Winner which is currently active there.

Severo-Varyoganskoe

Oil field in Russia having been producing for over 30 years. Ten expired wells at the field have recently been brought back into production by hydraulic fracturing. This has realised about 1000 barrels per day. The hydraulic fracturing was carried out by TNK-BP, who are also engaged in hydraulic fracturing at *Lugansk* in the Ukraine where a gas field is being developed. TNK-BP are also active at the oil field *Kamennoe* in western Siberia where again hydraulic fracture is being applied.

Shah Deniz

BP project at a condensate field in the Caspian Sea. A **TGP** 500 **jack-up** rig is in service there, as at the **Harding Field**. The rig was installed on a **template** for drilling in 105 m of water and will, as at the Harding Field, become a production facility once drilling is finished. A multiple **pig** launching system is in use there.

Shale natural gas

Such gas, from reservoirs composed of sedimentary rock, is becoming increasingly important in North America. The table below contains information on three selected shale natural gas fields, all operated by Chesapeake Energy. Comments follow the table.

Field and location	Details
Marcellus[58], extending into OH, NY, PA, WV.	**Pay** of 275 m in places. First production 2005. Now >100 production wells over an area of 3600 square kilometres. Permeabilities in the range 10^{-2} millidarcy down to 10^{-5} millidarcy.
Barnett, TX.	Discovered 1981. Producing 140 million cubic metres per day in 2010. Probable that the proposed LNG liquefaction plant at **Freeport TX** will be supplied from Barnett and from Haynesville (next row).
Haynesville, extending into TX, LA and AR.	Producing at 22 million cubic metres per day. Hydraulic fracture with sand as the **proppant**. Scene of use of **SHALE-DRIL™**, a Haliburton product, and the **Spear drill bit**.

At Marcellus the natural fractures ('joints') in the formation are vertical, so penetrating them requires horizontal drilling enabling gas so liberated to flow into the well bore. Horizontal drilling was carried out in such a way as to open up as many of the vertically orientated joints as possible. Note the very low permeability range. At Marcellus and at Barnett (next row) the gas is 'tight' in the sense of that term explained in other entries of this volume although the term 'tight gas field' is also applied to fields where the rock is non-sedimentary. If only one descriptor precedes 'natural gas', 'shale' is more immediately relevant than 'tight'[59]. Haynesville is also horizontally drilled. A spin-off point from the row for Haynesville is that hydraulic fracture can be applied to sedimentary rocks. The **hydraulic fracturing fluid** used at Haynesville is 99% sand and water. (*See also* **AziTrack™**; **Argentina, unconventional gas plays in**; **Australia, shale natural gas in**; **Blackpool, Lancashire**; **Changbei shale natural gas field**; **Horton Bluff shale natural gas**; **Karoo Basin**; **Poland, shale natural gas in**; **USA, selected CBM reserves in**; **Salah Gas**; **Utica natural gas shale field**.)

Shallow seawater, possible difficulties with

Over the 65 or so years of the offshore oil and gas industry deeper and deeper waters have been accessed and 'progress' has been judged by the depth of sea-floor reached. It must not however be forgotten that when the water happens to be shallow, say only of the order of tens of metres, that can make for difficulties if there is insufficient space between **moonpool** and sea-floor. That is where the **jack-up** rig comes into its own, a point sometimes made by builders of jack-up rigs. As noted new-build jack-ups are now quite numerous. Even so, a jack-up rig will have a minimum water depth for operation: that for **Lloyd Noble** is about 3.5 m. It is noted in the entry for **Stena Don** that it can drill only in sea depths in the range 130 m to 500 m. (*See also* **Sunkar floating production vessel; Troy Williams**.)

Shanghai LNG project

'Peak shaving' was the *raison d'etre* of LNG when it was first introduced in Cleveland OH in the 1940s. The meaning is simply that LNG would be held in reserve for evaporation and reticulation at periods of 'peak' demand. This practice is not obsolete, and in Shanghai, China an LNG terminal for this express purpose was opened in 2008. This marked the first phase of the Shanghai LNG Project, a receiving terminal for LNG from Bintulu in Malaysia. It is expected that its capacity will be 3 million tonnes per year by 2012 and 6 million tonnes per year by 2020 if not earlier. It receives LNG from Bintulu, Malaysia in tanker loads of up to 165 000 cubic metres. The facility has **SCV** and also **IFV**. (*See* **Weaver's Cove MA**.)

Shearwater Field

Condensate field in the North Sea in a water depth of 90 m. The operator is Shell. Gas and condensate are taken from a wellhead platform to an offshore processing facility 80 m away. The reservoir is as deep as 4545 m. The field has been producing since 2000, and drilling of the production wells had been challenging. Experience from drilling at the nearby Franklin Field, where a **PDC** drill bit had had to be in some degree specially fabricated in terms of weight and number of blades, was drawn on. At Shearwater such a bit achieved an **ROP** of 15.4 feet per hour in sandstone along part of the drilling trajectory.

Shelley Field

Small oil field in the UK sector of the North Sea, water depth 95 m, producing (from two wells) over the period 2009–2010. The operator was Premier Oil. The appraisal well, drilled by **Ocean Guardian** in 2007, was at a **TVD** of 4221 m and the oil from it was 31 degrees API (density

870 kg m^{-3}). Development drilling was by Sedco 712, a **semi-submersible rig** owned by Transocean, and production was by the **FPSO** Sevan Voyageur which used polyester mooring lines. Decommissioning was in 2010–2011 so the period of production was short. Such a course of events for a small field should not be taken to signify 'failure', especially at a time of fluctuating oil prices.

Shock absorber

The power to a drill string is very high. The resistance to drill bit rotation is also very high, that being in a sense the raison d'etre of the bit. That there will be vibrations in drill string operation is therefore obvious. Breakages and reductions in bit life and in **ROP** are the result, consequently shock absorbers are put in place. For the purposes of this entry one such device – the Dailey® R-A-M® Shock Absorber from Weatherford in TX – will be discussed in some detail. It comes in a range of sizes, from 6½ inches o.d. to 11 inches o.d. Considering the 9 inch o.d. model, it is 14.5 feet long and weighs 4000 lb (1.8 tonnes). It can function at a **WOB** up to 11000 lb (490 kN) and at temperatures up to 200 °C. For a device intended to absorb vibration the most important performance criterion is surely the dynamic spring rate, units N mm^{-1}. This quantity is also important in the matter of vehicle suspension. For a single elastomer substance it is not a 'physical constant' but depends on the strain being experienced. In shock absorber design and performance it is a characterising specification of the entire device. In a motor vehicle shock absorber the value will be tens or possible hundreds of N mm^{-1}. The Weatherford web pages give a calculated value for the 9 inch o.d. Dailey® R-A-M® Shock Absorber of 16 000 N mm^{-1}. When an industrial shock absorber initially responds to a 'shock' (perhaps more accurately a vibration: 'vibration damper' is probably a better term, but less widely used) there is movement of or within the shock absorber. The distance between initiation and cessation of movement is called the stroke, and is controlled by the stiffness. The author has obtained information on this for a manufacturer of oil drilling shock absorbers other than Dailey®R-A-M® which, like the latter, comes in a range of sizes[60]. The one of o.d. 160 mm (6.3 inches) has a length of 5.25 m and a maximum stroke of 120 mm (30.5 inches). The 6.3 inch o.d. model was chosen as an illustrative example because that has roughly the same o.d. as the Dailey®R-A-M® which was analysed above, but the comparison must not be taken any further. It might be noted however that the maximum WOB at which it can function is 340 kN, about 70% of the upper limit for the Dailey® R-A-M®. (*See also* **AVDTM**; **Shock Sub System**.)

Shock Sub System

In the entry for **shock absorber** generically the example chosen for analysis was that from the Dailey® R-A-M® Shock Absorber and the spring rate was given as 16 000 N mm^{-1}. The Shock Sub System manufactured by JA Oil Field Manufacturing Inc. (HQ in Oklahoma City), is a similar tool. Its top-of-the-range model has a spring rate of:

$$52000 \text{ lb per inch} = 9108 \text{ N mm}^{-1}$$

which is about 40% lower than that for the Dailey® R-A-M® Shock Absorber.

Shtokman gas condensate

This reserve of gas and condensate in the Barents Sea was discovered in 1988. Development is under way in the expectation that production will begin in 2016. The water depth is 50 m. Plans include the following. Four production platforms will be needed for the eventual targeted production, so **spar platform** and **tension leg platform** designs are being evaluated. Each platform will have an ice wall. The gas will be transported to Vyborg by a pipeline which will have as many as ten **compressor stations**. The sea-floor is very uneven and dredging will be needed in advance of pipe laying. Eventually there will be LNG trains at Vyborg; the LNG product will be for the US market whilst the NGL might be for the local market. Two **semi-submersible rigs** for well drilling are under construction in Vyborg. (*See* **Jeanne d'Arc Basin**.)

Shushufindi Field

Oil field in Ecuador having produced about a billion barrels. **Spudding in** was in 1968 and the depth of the first production well (in 1972) was 3000 m. Its production rate in 2010 was 85 000 barrels per day. In 2007 further exploration wells were drilled by the state oil company Petroecuador and a rise in the daily production rate to 100 000 barrels per day is hoped for. (*See* **Sacha 171H well**.)

Shuttle tanker

This is used to receive oil from an **FPSO** or an **FSO** in its 'O' – Offloading – role for transfer to shore. Condensate is also so transferred. Some shuttle tankers use **dynamic positioning**, others 'conventional positioning' such as has featured in other entries of this volume. A major manufacturer of shuttle tankers (amongst other types of vessel including FPSOs) is Teekay whose HQ is in Vancouver. Thirty-five Teekay shuttle tankers are currently in service. Brief details of a selection of them are given in the table below.

Name of vessel	Year of entering service	Quantity of oil which can be held	Further information
Aberdeen	1996	0.56 million bbl	Like the majority of Teekay vessels displays the flag of the Bahamas.
Basker Spirit	1992	0.68 million bbl	Built at the Dalian yard in China.
Navion Britannia	1998	0.86 million bbl	Built at a Spanish yard. See note on 'Navion' in the main text.
Navion Clipper	1993	0.51 million bbl	Built by Mitsui in Japan. About half of Teekay's fleet of shuttle tankers are called 'Navion …'.
Nordic Brasilia	2004	1 million bbl	Built by Samsung. Transfers oil produced in the Campos Basin to Sao Paulo.
Randgrid	1995	0.87 million bbl	In service off Norway.
Vigdis Knutsen	1993	0.86 million bbl	Displays the flag of the IoM.

'Navion' was the tanker division of the Norwegian oil company Statoil before it was sold to Teekay. Volatile organic compounds (VOC) are a major pollutant of the atmosphere. All of the shuttle tankers in the table above are fitted with vapour emission control to prevent release of VOC and they all have double hulls. A feature of any shuttle tanker will be cargo heating to prevent deposition of heavier components of the crude oil. The capacities (column three in the table) of shuttle tankers are such that they are suitable for conversion to **FSO**s and Teekay have in fact entered this line of business. The company's July 2009 newsletter reported that Rita Knutsen, a shuttle tanker acquired by Teekay in 2005 with modification in mind, is scheduled for conversion to an FSO for use offshore Qatar and that this will be the debut of Teekay's fleet in the Middle East.

The force from the engine has to act against the balance of weight of the shuttle tanker and the weight of water it displaces, which will vary continually but will only ever be a small proportion of either of the weights of which it is the difference. Using numerical values for the shuttle tanker Aberdeen (row one of the table) calculations concerning performance are attempted in the box below.

Adding the 'light ship weight' to the 'deadweight'[61] gives for the displacement a value:

$$(17\,400 + 87\,055)\ \mathrm{t} = 104\,455\ \mathrm{t}$$

At full speed the vessel travels at about 20 knots (10 m s^{-1})* with an engine providing 19 135 h.p. or 14 269 kW. Assigning the symbol $\boldsymbol{W}$ (units kg) to the balance of weight and buoyancy forces referred to:

$$(\boldsymbol{W}\ \mathrm{kg} \times 10\ \mathrm{m\ s^{-1}} \times 9.81\ \mathrm{m\ s^{-2}}) = 14\,269\,000\ \mathrm{W}$$

$$\downarrow$$

$$\boldsymbol{W} = 145\,000\ \mathrm{kg\ or}\ 145\ \mathrm{t}$$

and this is less than 1% of the shuttle tanker weight alone, and less than 0.1% of the weight of the shuttle tanker when it is loaded to the maximum immersion permitted. As stated, it represents the interplay of vessel weight and buoyancy forces.

*Author's estimate, being a generic value for the maximum speed of such tankers.

The calculation gives a perspective on power requirements in vessels. These, re-expressed as weights, will be much smaller than the vessel weight because of the effects of seawater displacement. One can also glean from these figures how a simple device such as a set of wires will hold a tanker stationary. The drift velocity will be much lower than the cruising speed above and the value of ***W*** correspondingly smaller, enabling suitably positioned metal or (as we have seen elsewhere) polyester lines to accommodate the force with ease. That the same lines could raise the tanker against gravity is quite inconceivable (*See also* **Glas Dowr**; **Glitne Field**; **Gullfaks**; **Platong II Project**.)

Shwe Field

Gas field offshore Mynamar, under development. There have been a number of appraisal wells. Participants in the project include Daewoo. It is part of the plan that some of the gas will go west to India and some north-east to China. (*See also* **Myanmar-China pipeline project**; **Nanhai II**.)

Side dumping

A.k.a. rock dumping, an alternative to use of fall pipe mounted at a **moon-pool** in a subsea rock installation. It has instead a moveable *skid* mounted on the edge of the deck of the vessel which will be tilted for rock admittance to the sea. An alternative is for gravel to be spread across the deck and transferred overboard by rotating blades, as with **Pompeii**. The **SSDV** fleet of the world is considerable and of course offshore oil production is not the only application. (*See also* **Cetus**; **GDF Suez Neptune**.)

Sidetracking, by coiled tubing drilling

It is explained in the entry for **TTRD** how sidetracking – formation of a branch from and at an angle to the well casing – can be achieved by rotary drilling and use of a **whipstock. Coiled tubing** drilling has also found wide application to sidetracking. The table below gives a selection of such operations.

Location	Details
Alwyn North Field, North Sea	Well 3/9a-N30 sidetracked by coiled tubing drilling. Whipstock set at a distance of 3380 m along the existing casing and coiled tubing drilling from there. Sidetrack now yielding 1100 barrels of crude oil per day.
House Mountain oil field, Alberta	Horizontal sidetrack using underbalanced coiled tubing drilling. 3⅞ inch section attached at 90 degrees to an existing well casing. A drill bit of the necessary diameter supported and supplied with fluid by a 2⅜ inch coiled tube.
Prudhoe Bay oil field, Alaska	Many wells given extended life by creation of sidetracks using coiled tubing drilling as well as by TTRD. See comments following the table.
Dalen 2 gas field, Netherlands	One of 12 wells selected for **UBCTD**. Sidetrack installed in the the casing of the well, the **TVD** of which was 2890 m. Sodium chloride brine used as **drilling fluid** to create underbalanced conditions of drilling. Coiled tubing of 2 inch diameter for supply.
Slaughter oil field, West Texas	Existing well of measured depth (MD) 1600 m. Sidetrack at 1475 m along the original casing by coiled tubing drilling

A whipstock can be either retrievable or non-retrievable. That used at Alwyn North was of the retrievable type. The flow of oil from the sidetrack at Alwyn North looks modest enough. At the Brent price of crude oil on the day that this entry is being written, one year's oil from the sidetrack is worth $US36.7 million. The important point is that this soon enabled the sidetracking operation to pay for itself, a point made in a coverage of the matter in the professional literature. It is also, as a bonus, obtaining further use from the original well casing. We note in passing that the Alwyn Field, where the formation is sandstone, has one of the highest permeabilities of all of the fields in the UK sector of the North Sea, approaching 1 darcy. At the Horse Mountain Field a non-retrievable whipstock (a.k.a. as a single trip whipstock) was used. Horse Mountain is an example of UBCTD. In 2010 BP carried out approximately 35 sidetracking projects with coiled tubing drilling at Prudhoe Bay and about half as many with rotary drilling. There had been more by each method the previous year. The Dalen Field (next row) is producing gas at 0.36 million cubic metres per day, a result lower than that hoped for by the promoters of the sidetrack project. The brine used in drilling had a specific gravity of 1.06 (equivalent to 8.83 **ppg**). It is quite possible for water alone to be used in UBCTD. (*See also* **Auk Field**; **Vesterled pipeline**.)

Sigyn Field

Gas field in the Norwegian sector of the North Sea. It is 12 km from the Sleipner Field, to which its production is tied back. The operator of the Sleipner Field is Statoil, who operate Sigyn but not exclusively. There is significant condensate accompanying the gas from Sigyn.

Silica gel, use of in natural gas dehydration

Other entries in this volume have been concerned with natural gas dehydration by two approaches: **glycol dehydration** and the use of molecular sieves. Silica gel is also so used, and is capable of bringing gas down to the 'pipeline standard' of circa 7 lb per million cubic feet of gas. (See also **GasDry**; **Sorbead®**.)

Silver Bullet

Range of **PDC** drill bits manufactured by Torquato (HQ in PA). Those used in oil and gas well drilling are five-blade or six-blade in configuration and can be steered in **directional drilling**. They have **tungsten carbide** inserts as part of the **matrix body** to ensure bore diameter consistency. In addition to multi-blade forms a flat face version is available and this has found application to **CBM** wells in small bit diameters (< 2 inches) clearly classifiable as **slimhole drilling**.

Simon Stevin

Newly constructed **DPFV**. It can accurately direct stones to sea depths of up to 2000 m. It can carry 33.5 tonnes of stones/rock for such use and it can travel at 15 knots with a full payload. The **POB** is up to 70.

Singleton Field

Onshore field in south-eastern England where production began in 1989. The operator is Providence Oil plc. There are seven production wells yielding in total 800 barrels per day and **pump jacks** are in operation. An additional well has been drilled directionally in order to access oil in a **fault** block. The **pay** so accessed will be as extensive as 700 m.

Sinphuhorm gas field

In Thailand, having Salamander amongst its stakeholders. The formation is limestone and dolomite. Production is 2.5 million cubic metres per day of gas and 500 barrels per day of condensate. The gas is used in electricity generation. (*See* **Bualuang Field**.)

Skandi Aker

Well intervention vessel built in Norway and operated by Aker Solutions. It can operate at well depths up to 3000 m and has drilling capability, unusual for a well intervention vessel. It has **dynamic positioning** and a crane capable of bearing 400 tonnes. Comparison, for which there is a reasonable basis, with for example the drill ship **Deepwater Champion** the **derrick** of which can take 1800 tonnes, reveals that in its drilling capability Skandi Aker is lightweight but that is by design, as it is seen as a well intervention vessel with limited drilling capability possibly to be deployed for sidetracking.

Skarv

Newly built **FPSO**. Its start-up in the field of the same name offshore Norway is taking place at the time of going to press. It has **turret mooring** and can store 0.875 million barrels of oil. Its processing capability is 0.42 million cubic metres of gas per day and 80 000 barrels of oil per day and initially it will receive from 16 production wells. Gas will be taken to Kårstø by pipeline and the oil transferred to shuttle tankers. (*See* **Gjøa Field**.)

Skua Field, North Sea[62]

In the **ETAP**, initially very productive. By 2004 product from the one well was deemed to be containing too much produced water for continued operation to be viable, that is, it was classified as a **watered out well**.

It is not however seen as having no future and development measures including **depletion drive** are being looked into by the operator Shell. (*See* **Marnock Field**.)

S-Lay

Means of installing offshore pipelines. Pipe is transferred from a vessel ('barge') to the sea via a device called a stinger, eventually reaching the sea-floor, its intended destination. Devices including a tensioning roller and one providing 'forward thrust' are applied during laying in order to prevent buckling of the pipe, thereby ensuring that it is horizontal when finally in situ. The profile of the pipe between vessel and sea-floor during laying forms a 'S' shape, hence the term S-Lay. Welding of one section of pipe to another takes place on the vessel before admittance to the stinger. The method can be used in water depths in excess of 1500 m at installation rates of about 4 miles of pipe per day. (*See also* **Acergy Condor**; **GSP Bigfoot 1**; **J-Lay**; **LB 200**; **Lorelay**; **Reel laying of pipelines**.)

Sledgehammer™

Drilling jar from Sperry Drilling Services. The detent force is six times the **WOB**, providing for vigorous movement. (*See* **Sup-R-Jar®**.)

Slickline

Wire used to lower a tool into a producing well or to retrieve such a tool after use from a well. Use of tools at wells currently producing is part of *well intervention* which can of course be on- or offshore. In the former case lowering or raising of slickline will be from a truck and in the latter case from a well intervention vessel or quite simply from a slickline lift boat such as **Troy Williams**. In either a **stuffing box**, which contains the well pressure in the event of slickline breakage, will be required. One of several major manufacturers of slickline is the Swedish company Sandvik. Their range varies in composition but, like all slicklines, have significant proportions of chromium, molybdenum and nickel. The diameter range is 0.082 inches to 0.150 inches (2.08 mm to 3.81 mm) and they are available in lengths up to 40 000 feet (12 000 m approx.) on a reel. Depending on composition and diameter, the breaking load of Sandvik slicklines is in the range 5 kN to 16 kN. The Sandvik product has a smooth surface which minimises removal of small amounts of wire by abrasion ('wire wool') on passage through the stuffing box. In the lowering and raising of a tool in a well a choice sometimes has to be made between slickline and coiled tubing. An intervention vessel such as **Well Enhancer** will be equipped with both. (*See* **Braided line**.)

Slimhole drilling

It has been explained how in **spudding in** of an oil well a drill bit of up to two feet in diameter is used, progressively smaller drill bits being used until the reservoir is reached. The final diameter is that at which **completion** commences and there is sometimes a strong incentive to make the diameter at this site as small as possible to minimise the extent of the installations and amounts of fluids at completion. When in drilling the casing diameter is kept below what would be usual for the depth with this end in view, the term slimhole drilling applies. Advantages of slimhole drilling include smaller volumes of well cuttings. One arbitrary definition of slimhole (amongst others) is a bore of diameter six inches or less for 90% of the drilled length. If a six-inch bore can be drilled instead of an eighteen-inch one to a depth of 4500 m the amount of cuttings will be reduced by 65% to 70%. There are also savings on casing as well as on completion as noted, and economy is the usual reason for choice of slimhole drilling. (*See also* **Xplorer**; **Fort Yukon AK, CBM at**.)

Snøhvit LNG facility

This takes natural gas from the Albatross and Askeladd fields offshore Norway (sea depth 250 m to 345 m) and converts them to LNG. Pipelines from **Melkøya Island** to the Snøhvit Field were laid by **Solitaire** over about a two-month period in 2005. Liquefaction takes place on a vessel ('liquefaction barge'). One reason for this was to reduce the amount and weight of metal hardware on Melkøya Island. Liquefaction at the barge uses the spiral wound heat exchanger (SWHE). (*See also* **FNGU**; **Liquefied natural gas (LNG), manufacture of**.)

Snorre Field

In the Norwegian sector of the North Sea at a water depth of 300 m to 350 m. Production began in 1992 by use of a **tension leg platform** (Snorre TLP). That was in the southern part of the field, and a semi-submersible production facility has been in use at the northern part since 2001. Both production facilities are still in use. Well depths at Snorre are 2500 m to 3000 m and some are horizontal. There is also **water injection** and gas reinjection.

Sodegaura LNG terminal

Japan imports a large amount of LNG and, therefore, needs terminals at which to collect, store and regasify it. Sodeguara is one three such facilities operated by Tokyo Gas. Said the be the biggest LNG terminal in the world, Sodegaura commenced operations in 1973 and has received LNG from places including Alaska, Australia, Brunei, Malaysia, and Indonesia.

It is set up for **SCV** as well as having two **ORV**s. Negishi LNG terminal, also operated by Tokyo Gas (jointly with the Tokyo Electric Power Company, TEPCO) received its first shipment in November 1969 and with a shipment in November 2008 attained a cumulative amount of 100 million tons. A simple calculation follows.

> Averaging over the period and approximating tons to tonnes for simplicity, this terminal performance expressed in thermal terms is:
>
> $$100 \times 10^6 \times 10^3 \text{ kg} \times 55 \times 10^6 \text{ J kg}^{-1}/(19 \times 365 \times 24 \times 3600) \text{ s} = 9 \text{ GW}$$
>
> TEPCO use a range of fuels including coal and fuel oil. At Sodegaura they operate a 1000 MW power station with LNG (only) as fuel. The LNG requirement must be of the order of:
>
> 1000/0.35 MW = 2900 MW approx., 2.9 GW or about 30% of the averaged figure for the period of operation of the terminal.

and it is both interesting and reassuring when such basic calculations 'hang together'. The other LNG terminal operated by Tokyo Gas is at Ohgishima, where operations commenced in 1998. (*See also* **Dabhol LNG terminal**; **Kenai LNG plant**; **Lumut**.)

Sognefjord

Production well at the **Brage Field** which, interestingly, combines in a single well many of the production aid methods described in this volume. It is a horizontal well at 2000 m **TVD**. The permeability of the formation varies[63] and is as high as 1 darcy to 2 darcys in places. There is **depletion drive**, and the presence of a **gas cap** has eliminated any need for **water injection**.

Solan

Discovery in the North Sea west of Shetland, believed to contain 20 million barrels of oil. Drilling was by **Byford Dolphin** and when production commences it will be at a compliant tower, the first such in UK waters. Transfer of oil from Solan to **Schiehallion** is being considered: alternatives will be the use of **shuttle tanker**s taking oil from Solan directly *or* sharing of infrastructure with the **Foinaven Field**. (*See also* **Benguela Belize**; **Laggan and Tormore fields**.)

Solitaire

The largest **pipe laying vessel** in the world, she can carry 22 000 tonnes of pipe, enabling operation in the most hostile and remote areas without relying on frequent consignments of pipe transported by supply vessels.

Operational since 1998, Solitaire manoeuvres on full **dynamic positioning** and consistently achieves lay rates of 9 km (5.6 miles) per day. The vessel has set countless records for both lay speed and depth of pipelaying, most recently 2775 m. Solitaire can lay pipe with an O.D. range of 2 to 60 inches and operates five welding stations simultaneously. (*See also* **Audacia**; **Corrib Field**; **Fusion Bonded Epoxy (FBE)**; **LB 200**; **Snøhvit LNG facility**.)

Songkhla oil field

In the Gulf of Thailand, productive since 2009. The water is very shallow, < 25 m. Several wells have been drilled, and the Songkhla A-09 well has a **TVD** of 2470 m and the **pay** is 65 m. **Spudding in** has taken place at Songkhla A-12. The shallow depth makes for straightforward operation of a production vessel, from which the oil is transferred to tankers. Current production is 10 000 barrels per day.

Sonication, well stimulation by

Research into well stimulation by **acoustic excitation** is taking place. Its benefits if it becomes commercialised will include a gentler treatment of the formation than with the use of chemical agents, especially important close to the **BHA**. Although the R&D is in the US, certain Nigerian fields have been identified as potential beneficiaries.

Sorbead®

Drying agent for natural gas, composition 97% silica, 3% alumina. Its BET surface area is as high as 650 $m^2\ g^{-1}$. In 2009 BASF undertook to supply 2000 tonnes of Sorbead® for use at the new Portovaya **compressor station** in Russia. This will incorporate the largest silica gel gas drying plant in the world, so treating 170 million cubic metres of gas per day. Related calculations follow.

Noting the surface area of 650 $m^2\ g^{-1}$ and calculating the area of one water molecule as:

$$[(1/1000)\ m^3\ kg^{-1} \times 0.018\ kg\ mol^{-1}/6 \times 10^{23}\ mol^{-1}]^{2/3}\ m^2 = 10^{-19}\ m^2$$

If the water forms a monolayer the uptake per gram of Sorbead® is:

$$(650\ m^2 / 10^{-19}\ m^2)/6 \times 10^{23}] \times 18\ g = 0.2\ g.$$

That a sicila gel desiccant for natural gas will, depending on the quantity of water in the gas, take up to 15 to 40% of its own weight is a general result.

It was noted in the entry for **Glycol dehydration** that the usual industry standard of 7lb moisture per million cubic feet of gas sometimes has to be improved upon. How much Sorbead® or other silica gel desiccant would be needed to bring one million cubic metres of natural gas from the industry standard moisture content of 7lb per million cubic feet by half an order of magnitude?

Required removal of water = [7 – (7/5)] lb per million cubic feet
= 5.6 lb per million cubic feet

This is equivalent to 90kg per million cubic metres or 90 mg desiccant per cubic metre of gas. The required weight of desiccant is up to five times this, so up a quarter to a half of a gram.

Sorochinsko-Nikolskoe (S-N) Field

A Russian field in the Province of Orenburg where there are in fact a number of oil fields. At present several of them including S-N are in decline and 'rehabilitation' is under way. At S-N this takes to form of horizontal sidetracking of selected wells. **MWD** took place during sidetracking, in particular resistivity measurements. The target of the rehabilitation is an extra 1 million barrels over a five-year period. In Orenburg there is also an oil import terminal which receives from Kazakhstan. (*See* **Karachaganak Field**.)

Sour sealed truck

Hydrocarbon which is 'sour' – high in sulphur compounds including H_2S – will need to be desulphurised before refining. If however transport to a refinery is not by pipeline but by road truck, precautions are needed to ensure that the truck along its route does not release sulphur compounds. A truck incorporating such precautions is called a sour sealed truck and manufacturers of such trucks include Peak Energy, who manufacture specialised trucks for the oil industry across the western US and Canada. In a sour sealed truck the oil is admitted to a tank, supported on a chassis, which has been partially evacuated, and its contents remain below atmospheric pressure after filling, precluding oil vapour exit. However it is recognised that there might be contingencies such as a highway accident when it is necessary to vent the truck to the atmosphere, and a single-use desulphuriser is present in the tank for that purpose.

South Arne Field

Oil field in the Danish sector of the North Sea, water depth 60 m. The operator is Amerada Hess. There are currently five producing wells yielding a total 50 000 barrels per day. The field is undergoing expansion which involves hydraulic fracture with sand as the **proppant**. (*See also* **South Tor Pod**; **Tyra Field**.)

South-East to North-East pipeline

In Brazil, having been in use since September 2010 and owned and operated by Petrobras. Having a total length of 1371 km, it carries gas originating from fields in the Campos Basin. Its current load is 10 million cubic metres of gas per day, to be doubled by 2015 by the installation of **compressor stations**. Elsewhere in this volume[64] it is stated:

> *the power required for gas compressor operation is a thousandth the hypothetical heat release rate for burning the gas at the compression rate*

and on this basis it ought to be possible to make a prediction of the rating of the compressor station that will be required to effect the increase. This is attempted in the calculation in the box below.

$$[20 \times 10^6\ \mathrm{m}^3 \times 37 \times 10^6\ \mathrm{J\ m}^{-3}/(24 \times 3600)\ \mathrm{s}] = 9\ \mathrm{MW}\ (12000\ \mathrm{h.p.})$$

(*See also* **Pumping station**; **Rio de Janeiro-Belo Horizonte Gas Pipeline II**.)

South Furious Field

Offshore Malaysia, single well of **TVD** 660 m and measured depth (MD) 875 m. Very much a 'marginal field', its viability might have been enhanced by the fact that it was one of the first to have **zonal isolation** installed.

South Tor Pod

Five kilometres south-east of the **South Arne Field**, and the scene of exploration. The formation is chalk, and drilling of an exploration well was by **Ensco 70**, at a water depth of 55 m and to a subsea depth of 2968 m. There was also a sidetrack from the vertical production well.

Southern Natural Gas Company

Based in Birmingham AL and operator of the Southern Natural pipeline, which originates in Louisiana and terminates in Georgia, therefore conveying gas in an easterly direction. It has 40 **compressor stations** rated jointly at 566 280 h.p. and can carry 870 million cubic metres of gas per day.

Southern North Sea, tight gas in

It is stated in the entry **North Sea, first hydraulic fracture operation in** that at the Beryl Field, where there has been hydraulic fracture to promote oil production, the permeability is about 10 millidarcy. There are gas fields in the southern North Sea where the permeability is in the range 0.5 millidarcy to 0.01 millidarcy and gas in such a field is 'tight gas'. In the southern North Sea fields, additional to the issue of tightness is the fact that in the geological formation the **pay** is intermittent. A rule of thumb in the industry is that hydrocarbon release without any sort of 'help' requires a permeability of 100 millidarcy[65]. The bottom-of-the-range figure for the North Sea fields given above is four orders of magnitude lower than this, and this gives a perspective on the challenges involved in accessing such gas. The way forward *if* the gas is to be made available for use is hydraulic fracturing. Experience in the North Sea generally is being drawn on in the **Cardiff (NZ) development**. (*See also* **Angot Field**; **Basin centred gas**; **Cusiana Field**; **Shale natural gas**; **Sidetracking, by coiled tubing drilling**; **Tight gas, South American sources of**; **Tyra Field**.)

Spar platform

In this configuration an offshore production facility is mounted on a cylindrical hull with its axis perpendicular to the surface of the sea. The cylinder becomes stationary on a simple Archimedean basis when the weight of water displaced equals that of the weight of the hull plus its load (the 'topside') and is held in position by wires and chains to the sea-floor. Typical dimensions for the hull of a spar platform in the Gulf of Mexico are diameter 4 m, height 250 m. Immersion will be about 90%, so if the precise value of the height was 250 m only something like 25 m would be above the sea. Sea depths of up to 3000 m can be accessed with a spar platform. There is a difference in the physics of the spar platform and the **tension leg platform**. In the latter tension is introduced into the 'tendons' which remain taut. In the former the wires and chains show catenary behaviour; they are influenced only by their own weight and their contact with seawater and are not under high tension. The assembly of chains and wires in this form provides for countermovement as required to any drift caused by the sea. There is limited basis for comparison of the operation of a spar platform and **dynamic positioning**. The spar arrangement can be used in drilling as well as in production as at **Kikeh**. Metals wires, ropes, chains and leads for use in the sea can be protected from corrosion by being galvanised. A spar will sometimes have drilling facilities as well as production ones, largely because of the possible need for **water injection** wells as production takes its course. (*See also* **Boomvang**; **Constitution/Ticonderoga Field**; **Perdido platform**.)

Spear drill bit

Schlumberger product, having a steel body and a **PDC** cutting surface. It was fabricated expressly for use at the **shale natural gas** deposit at Haynesville. There it was required that drilling along a curved trajectory with intervals of lateral drilling was accomplished in one drill run, without the need change the drill bit or to raise it and reorientate it. In bend drilling the degree of curvature was 8 degrees to 14 degrees per 100 feet drilled. Over the entire drilling distance – curved and lateral – of 6063 feet an average **ROP** of 49.7 feet per hour was attained.

Spread mooring

Term applied in **FSO**s and **FPSO**s to mooring by means of lines to the sea-floor from several points on the structure being moored. By analogy with single point mooring it could be called 'multi-point mooring'. With **turret mooring** the 'weather vaning' is allowed for as a way of providing flexibility of positioning. With spread mooring weather vaning is an unwanted effect and it is prevented partly by symmetrical configuration of the mooring lines. Spread mooring has been applied to good effect in FPSO operation at **Kizomba**. An FSO with spread mooring is in use at the Corocoro field offshore Venezuela. This is in shallow water (28 m) and has ten steel mooring lines. Weather vaning, precluded by spread mooring, is possible with **tower mooring**. (*See* **Knock Adoon**, **SARLAM**.)

Spud can

Support on the seabed for the leg of a **jack-up** rig. Obviously the legs of a jack-up rig will penetrate the seabed to a smaller degree with the spud cans in position than they would in their absence, but the spud cans themselves do enter the seabed to a depth of the order of metres. Their removal when the jack-up is dismantled for movement and use elsewhere is then required and this is called 'pull-out'. It is not at all uncommon for a jack-up rig to be returned to a scene where it has previously been in service. In that event the 'footprints' left by the spud cans can have effects extending beyond their own boundaries, that is, beyond the area they previously occupied and such effects can destabilise newly installed spud cans. As one example of the dimensions and construction of the spud can we can consider a set of four recently constructed by Lamprell for use with jack-up rigs in Norwegian waters. Each of the four will weigh 650 tonnes and have a diameter of 21 metres[66]. When a jack-up rig is dry docked, spud can inspection and repair is part of the programme, as at ASRY (Arab Ship Repair Yard) in Bahrain where jack-up rigs in the ENSCO fleet are sometimes 'hosted'. (*See also* **Atwood Aurora**; **GSF Adriatic VI**; **Mat support**; **Offshore Freedom**.)

Spudding in

The process of creating an exploration well begins with drilling using a drill bit of wide diameter, between about nine inches and two feet or more. The drill bit will usually have either **tungsten carbide** or diamond as the cutting material. As drilling takes its course and the well deepens, successively smaller bits are used. A well having received the initial treatment with a wide drill bit is said to have been 'spudded'. **Drilling fluid** at the spudding in stage is likely to comprise **bentonite** and water only, and is termed 'spud mud'. Information for a spud mud is given when **non-Newtonian fluid** behaviour of drilling muds is discussed. (*See also* **Sangaw North**; **Stella development** inter alia.)

SRB

Sulphate reducing bacteria. These cause difficulties in oil pipelines. They oxidise hydrocarbons and in so doing cause a change in oxidation number of the sulphur from +6 to −2. The reduced sulphur becomes hydrogen sulphide and/or a sulphide ion, either of which is unwelcome in an oil pipeline. A simplified account of the biochemistry will follow. Respiration is of course oxidation, which means acceptance of, as opposed to removal of, electrons. In almost all biological processes the acceptance of electrons by oxygen is a process which can be exemplified thus:

$$O_2 + 4H^+ + 4e \rightarrow 2H_2O$$

Oxygen in oxidation state 0 (↑ O_2) ***Oxygen in oxidation state −2*** (↑ H_2O)

With sulphate reducing bacteria the chemical species which accepts electrons is the sulphate ion SO_4^{2-}. The statement which is sometimes made that SRB 'breathe sulphate' therefore has a perfectly sound basis in biochemistry and is not a mere figure of speech. This can be exemplified, by analogy with the above:

$$SO_4^{2-} + 10H^+ + 8e \rightarrow H_2S + 4H_2O$$

Sulphur in oxidation state +6 (↑ SO_4^{2-}) ***Sulphur in oxidation state -2*** (↑ H_2S)

where electrons *accepted* by the sulphate are *donated* by hydrocarbons in the oil which are therefore themselves oxidised to such products as aldehydes and carboxylic acids. Sulphate ions can be present in seawater at up to 3000 p.p.m. and it is known from direct examination that SRB occur in the North Sea close to the platforms. Hydrogen sulphide is a toxic contaminant. Moreover, the above chemical equation could be re-expressed:

$$SO_4^{2-} + 8H^+ + 8e \rightarrow S^{2-} + 4H_2O$$

and formation of ions will be followed by corrosive activity possibly in synergism with **APB**. There are a number of genera of sulphate reducing bacteria. The most important in terms of oil pipelines include *Desulphovibrio, Desulphobulbus, Desulphomonas* and *Desulphonema.* SRB control is an issue at reservoirs as well as in pipelines, in particular when there has been **water injection**. The second of the boxed reaction schemes above can lead to hydrogen sulphide formation with a souring of the products from the reservoir. Control of 'upstream SRB' might feature as part of well stimulation. (*See also* **AMA®-324**; **BIOCIDE RX-1225**; **Bio-Clear™ 242D**; **Dagang Field**; **DBNPA**; **Desulphurisation of crude oil by SRB**; **MAGNACIDE® 575**; **Double hull tankers, SRB in**; **Hawtah Field**; **Hydraulic fracturing fluid**; **Niger Delta, SRB in**; **Nitrate ions, action on by SRB**.)

SRV

Shuttle and Regasification Vessel. (*See* **GDF Suez Neptune**.)

SSDV[67]

Side Stone Dumping Vessel. (*See* **Side dumping**.)

SSMA

Sulfonated polystyrene-maleic anhydride copolymer, sometimes an ingredient of water based drilling muds. In drilling muds containing **bentonite** it stabilises the mud by neutralising charges having developed on the particles thereby preventing their coalescence. It is also used in water based muds with polymer viscosifiers. For example the Baker Hughes products ALL-TEMP™ and MILL-TEMP™ have been developed for addition to such muds where they have the function of preventing coagulation and deposition of the viscosifying polymer.

Stabilised condensate

Natural gas condensate having undergone refluxing to bring its Reid Vapour Pressure (RVP) down to a value ensuring safety in storage and, more especially, in tanker transportation. Target RVP values will usually be in the 10 p.s.i. to 12 p.s.i. range. Usually a stabilised condensate will have an octane number of only about 60 to 70 and cannot be used in spark ignition engines without octane enhancement. (*See also* **Condensate stripping**; **Jebel Ali Refinery**; **Kårstø processing plant**; **Kharg Island, Iran**; **Platong II Project**; **Torm Ugland**.)

StarBurst™

Whipstock from Weatherford, recently put to very effective use in a field in the Norwegian sector of the North Sea operated by Statoil. In an exploration well there it formed a diversion of diameter 8½ inches at an inclination of approximately 68 degrees at a well depth of 4136 m. In exploratory drilling, rig time is a major factor in cost, and time saved by measures such as that described in this entry can make the difference between feasibility or otherwise of a particular exploration. It is noted in Weatherford's coverage of the matter that 'rig rates' are very high at the place where their successful whipstock operation was carried out.

Statfjord Field

Oil field, like the **Blane Field**, straddling the UK and Norwegian sectors of the North Sea. The water depth is 150 m and the **pay** is at depths in the formation in the range 2500 m to 3000 m. It was discovered in 1974 and began producing in 1979. Oil and associated gas from Statfjord go to the **Kårstø processing plant**. There are three production platforms at the field, Statfjord A, Statfjord B and Statfjord C, each a **GBS**. (*See* **Nitrate ions, action on by SRB**.)

Statpipe

Natural gas pipeline, linking gas from fields in the Norwegian sector of the North Sea (including **Gullfaks**) to sales destinations in mainland Europe. It was opened in October 1985 and has conveyed 120 billion cubic metres of gas since, a heat supply averaging to:

$$[120 \times 10^9 \text{ m}^3 \times 37 \times 10^6 \text{ J m}^{-3}/(25 \times 365 \times 24 \times 3600)] \text{ W} = 6 \text{ GW approx.}$$

The pipe is of total length 880 km, 25% shorter than the much more recently laid Langaled pipeline, which conveys natural gas from Norway to England. (*See also* **Kårstø processing plant**; **LB 200**, **Zeepipe I**.)

STEELSEAL®

'**Lost circulation** material', a Halliburton product. It is obtainable with different diameter ranges of the carbon particles it contains, roughly from 100 μm to 1 mm. The carbon particles enter spaces in the well formation and expand, thereby preventing the lost circulation which the spaces would otherwise have caused. A recent 'success story' of STEELSEAL® was at **Thamama**.

Stella development

Scene of a gas and oil field in the North Sea, discovered in 1979 but only now undergoing major work. **Spudding in** of an appraisal well took place in April 2010 using a **jack-up** rig. The **pay** was found to exceed 250 m.

Stena Don

Semi-submersible rig currently in service offshore Greenland where it is being used in the drilling of exploratory wells. It can drill at sea depths in the range 130 m to 500 m and has **dynamic positioning**. The maximum **POB** is 102. (*See* **Shallow seawater, possible difficulties with**.)

Stena Forth

Drill ship currently in service offshore Greenland. It can drill at water depths up to 3050 m. It has **dynamic positioning** and a maximum **POB** of 180. (*See* **Disko Island**.)

Stena Spey

Semi-submersible rig having done major service in the North Sea. After work in the **Strathspey Field** and in the **Captain Field** its topside became contaminated with oil water and the absorbent product CrudeSorb® was used to very good effect. CrudeSorb® was developed and is made by the same company who produce the **CrudeSep®** device.

Stim Star

Family of **well stimulation vessels** operated by Halliburton. One of the most recent is Stim Star Borneo. Its features include the ability to store up to 160 tonnes of **proppant** for hydraulic fracture, and up to 5200 gallons of acid for **matrix acidising**. **Hydraulic fracturing fluid** can be transferred from vessel to well at a rate of up to 25 barrels per minute. A sister vessel is Stim Star II, which is in service in the Gulf of Mexico. A précis of one of its success stories follows. At a well at an offshore non-associated gas field, **Frac-Pack** treatment was carried out by Stim Star II. The hydraulic fracturing fluid was Delta Frac®, a Halliburton product containing borate. Production at the well was raised by a factor of 2.5. (See **San Jaun Basin NM**.)

STL

Submerged turret loading. (*See* **Fulmar Field**.)

STP

Submerged turret production. (*See also* **Yùum K'ak' Náab**; **Raroa**.)

Strathspey Field

Oil field in the North Sea, water depth 135 m. Oil from it is taken to a platform at the nearby Ninian Field. In the terminology of offshore engineering it is *tied back* to Ninian. Strathspey is operated by Texaco and Ninian by Canadian Natural Resources (CNR), so the effect of the tie-back is to divert the oil to the custody of an operator other than that of the field of origin. This was the first such commercial arrangement in the North Sea. (*See also* **Blane Field**; **Rhum Field**.)

Stuffing box

In well engineering, a means of containing the well pressure during **slickline** operations if breakage of the slickline occurs. There are many available, and for illustrative purposes we examine the range offered by Lee Specialities in Alberta. This range takes in pressures from 5000 p.s.i. to 15 000 p.s.i., the respective weights of the units being 13.6 kg and 47.6 kg. Lee Specialties can supply a sheave to match, through which slickline passes before returning to the main reel. This is a common arrangement in slickline operations. At onshore fields there will often be an appliance for onshore well intervention which has both a winch for release of slickline and a crane for lifting and positioning a stuffing box.

Subsea rock installation (*See* **Malampaya Field**.)

Suction anchor

Used with floating production facilities in deep water, a suction anchor is partially embedded in the sea-floor. Such embedment makes for greater effectiveness than the anchor's weight alone. (*See* **Girrasol**.)

Sugarkane Field

Gas and condensate field in Texas, discovered in 2006. The formation is carbonate. It was announced in August 2010 that the second of eight proposed wells at Sugar*loaf*, one area within Sugar*kane*, had been drilled to a depth of 1760 m. The operator at Sugarloaf is Hilcorp Energy Inc., HQ in Houston.

Sujawal X-1

Gas exploration well in the Sindh Province of Pakistan, depth 3000 metres **BKB**. There are significant amounts of condensate. The company carrying out the exploration is Mari Gas Company, Karachi, which about ten years after the separation of Pakistan from India began operations as Esso Eastern Inc.

Sulawesi

This region of Indonesia has been the scene of oil drilling since 1902, at the time the then Dutch East Indies were first taking an interest in oil: they later became exporters to Japan. Much more recently – within the last two to three years – ExxonMobil have been drilling at two blocks offshore Sulawesi. The decision was made partly on account of the company's need to close its operations in **Aceh**, and partly in response to the fall in oil production in Indonesia and the need to develop further reserves. The blocks where the drilling is taking place are called Mandar and Surumana and ExxonMobil in 2009 sold a 20% stake in each to Petronas. Both blocks are in deep water, 2000 m to 2200 m. (*See* **Bengkulu Basin**.)

Sulige Field

Tight gas field in inner Mongolia, scheduled for development by China National Petroleum Corporation and Total (France). Drilling is expected to commence in 2013. Total is also seeking partners for development of **shale natural gas** resource at Montelimar in France in the awareness that hydraulic fracture and horizontal wells will be needed for production. (*See* **Argentina, unconventional gas plays in**.)

Sunkar[68] floating production vessel

In service at the **Kashagan Field**, this vessel is not according to accepted terminology a 'production vessel' at all but a drill ship. It was converted expressly for its duties at Kashagan, which are in very shallow water but at very considerable well depths. The **POB** is 96. The rig is not moored – the water is too shallow – but has been placed on four limestone shelves ('berms') which are attached to the sea-floor by piles. These are below sea level at a water depth of about 4 m, and are themselves only about a metre in height. The piles are of 30 m depth and enable Sunkar to withstand ice impact.

Sunrise Field

Gas field offshore Australia in the Timor Sea, production expected to begin in 2013. Options for transfer of the gas to centres of population once it has been produced are being considered. One is a pipeline to Darwin. Another is conversion to liquefied natural gas (LNG) at a floating facility for such conversion. There are also major amounts of condensate at Sunrise.

Super Dry 2000/2000S

Flow assurance products from Weatherford, applied during operation of a **utility pig** so as to remove water before oil readmittance.

Sup-R-Jar®

Drilling jars from Schlumberger, in a selection of sizes. The one at mid-range (6½ inches o.d.) is capable of a detent force of 0.8 MN. This figure is further examined in the box below.

> The inevitable dangers of arguing from the particular to the general (or vice versa) are present here. However **WOB** values are often in the range 5 tons to 10 tons.

Using a value of 7.5 tons and equating tons to tonnes, the WOB becomes: 0.07 MN.

The detent force is therefore about an order of magnitude higher than the WOB. This is a sensible result, although it must be remembered that the detent force is applied only in a contingency and that the drilling jar is for almost all of the time just a passive part of the drill string.

The result calculated above is consistent with the same information for detent force and WOB provided by the manufacturers of **Sledgehammer™**, and not estimated as in the calculation above.

SureFire™

(See **Perforating gun**.*)*

SureGL™

Range of flow assurance products from Halliburton. They are admitted to a pipeline not as a plug as with a **gel pig** but in high-viscosity flow. (*See* **Black powder**.)

Sureshot™

Gamma detector for use in **MWD**, from APS Technology. Calibrated in API units, its measurement range is up to 800 such units and it can be used at well bore temperatures up to 175 °C. It is intended for concurrent use with other APS products, for example a resistivity monitor, in a modular arrangement. Power can be provided by a mud turbine or by batteries.

Su Tu Den

Name meaning 'Back Lion', oil field offshore Vietnam. An **FPSO** of the same name is in service at the field, at a water depth of 46 m. This was a new build. It has **turret mooring** and can hold one million barrels of oil. The operator of the field is PetroVietnam Oil Company (PV) which is owned by the government of Vietnam. Production in mid-2010, at which stage not all of the wells at the field had been put into operation, was 95 000 barrels per day. At that time PV sold a considerable quantity of oil to Shell, who themselves have no upstream activity at Su Tu Den. `There are a number of other fields close by (in the same block or an adjacent one), brief descriptions of which are given in the table below.

Field name	Details
Su Tu Nau ('Brown Lion')	North of Su Tu Den, undergoing development by PV which had previously been delayed since the field's discovery in 2005. Production in 2012 aimed for.

Field name	Details
Su Tu Vang ('Golden Lion')	20 000 barrels of oil per day being produced from 10 production wells. Oil taken to a central processing facility which also receives from Su Tu Den, and from there to an **FSO** of capacity 1.1 million barrels.
Su Tu Trang ('White Lion')	Condensate field, production hoped for by 2014. Plans are that the gas will go to one or more LNG trains.

(*See also* **Nam Con Son pipeline**; **Rong Doi Field**.)

Swellable elastomer

A means of reducing produced water, thereby increasing oil yield from a well. Applied to the outside of a casing and therefore annular in its orientation, the elastomer swells in response to water uptake thereby closing off sources of water. An elastomer for such use does *not* swell on contact with oil, so its effect is to close the water sources whilst leaving the part of the formation containing the **pay** open. There is so to speak an inversion of this method. To coat the outside of the casing with an elastomer which swells on uptake of *oil* will stabilise the casing and eliminate the need for cementing if oil is introduced as oil based drilling mud. This is probably the more common application having been applied for example at the Cormorant Field in the North Sea. There are some swellable elastomers which respond (i.e., by swelling) to oil or water and these can be used in lieu of cement with water based or oil based **drilling fluids**. Such an elastomer, capable of functioning at up to 10 000 p.s.i. and 120 °C, has recently been introduced onto the market by a company based in Aberdeen. One could deduce from the summary in this entry that the swellable elastomer has reservoir and well applications. Moreover, **zonal isolation** is often carried out at the site of **completion**. It is usually possible from logging for the location of water-producing lengths along the perforated tube at a completion to be known. The device used to block them so that the water is excluded (i.e. to achieve zonal isolation) is called a *packer* and use of a swellable elastomer is one of several possible approaches to this. (*See also* **Barton Field**; **Easywell**; **South Furious Field**.)

SWHP

Surface wellhead platform, whether a **fixed platform**, a **spar platform** or a **tension leg platform**. The term can be used in conjunction with other acronyms, for example one of the facilities at **Kizomba** has been assigned the name SWHP-TLP: surface wellhead platform – tension leg platform. (*See* **Peregrino Field**.)

Swinoujscie LNG Gas Terminal

In Poland on the Baltic coast, scheduled for commencement of operations in 2014. An incentive in its planning was reduction of natural gas supply from Russia and its replacement with LNG from countries including Norway, Denmark and Germany. Swinoujscie is seen as a more suitable venue for tankers from those countries than Gdansk would have been.

The world's largest pipe-laying vessel Solitaire *(courtesy Allseas)*

T

Tahiti Field

Site of a **spar platform** in the Gulf of Mexico, productive of oil and gas, at a water depth 1280 m. **Frac-Pack** well **completions** there were, at the time they were carried out, the deepest such completions, being at a well depth in excess of 7500 m. The operator is Chevron. Catenary lines are made from fixed lengths of chain purchased as such and joined together by shackles to the required length. There was delay in commencement of production at the Tahiti Field when the shackles initially obtained were found to be unsatisfactory. In September 2010 work began on the drilling of a **water injection** well at the Tahiti Field, the first deep-sea drilling in the Gulf of Mexico after the moratorium imposed as a result of the spill from the Macondo Field.

Tahoe

In the part of the Gulf of Mexico known as Viosca Knoll, the water depth is 457 m. Production of oil and gas began in 1994 and was tied back to Shell's Bud Platform some 20 km distant. Production was extended in 1996 by reason of four new production wells going to a new platform called Bud Lite, a **fixed platform** located in shallower water (84 m). Drilling for production expansion was from the **semi-submersible rig** Ocean Concord, part of the Diamond Offshore fleet, and the wells so created are all horizontal.

Talco Field

Oil field in Texas, discovered in 1936. After 30+ years of production from a well depth of 1280 m, its yield was greatly enhanced by the installation of an **electric submersible pump** in 1974.

Taldinskaya CBM field

In the Kemerovo region in western Siberia, recently the scene of the entry of Gazprom into **CBM** production. During trials in 2010 the company produced 5.3 million cubic metres of gas at Taldinskaya. Intervals of **pay** were found in the approximate depth range 330 m to 530 m. It is now intended that fourteen production wells will be drilled and that their gas will be used in power generation for the region, which is heavily industrialised. There are several other 'basins' in Russia known to contain CBM in large amounts which are being evaluated by Gazprom for possible production. (*See also* **Kovytka Field**; **Sekayu**; **Urengoy Field**.)

Tamar Field

Offshore Israel, under development in 2009 by Noble Energy, who have stated that it is the largest discovery that the company has ever made. It is expected to commence production in 2012. Drilling of the two appraisal wells was by **Atwood Hunter**. Tamar-I, the first appraisal well, is 90 miles out to sea from Haifa. The water depth is 1676 m and the well depth 3223 m. The corresponding figures for Tamar-II, 3.5 km from Tamar-I, are similar. The **pay** of gas is 140 m. (*See also* **Mari-B Field**; **Leviathan Field**; **Noa gas field**.)

Tanguh LNG plant

Whether Tanguh is in Indonesia or in PNG (as it certainly was at one time: it is on the island of New Guinea) has been the subject of legal deliberation and judgment, and it *is* now part of Indonesia. The LNG plant there came into operation in 2009, with gas from the Tangguh fields. There are two LNG trains with a combined capacity of seven million tonnes of LNG per year. The reserves of the field from which the gas is taken are such that up to eight LNG trains might, depending on the demand for, and price of, LNG, eventually be in service. There is of course also condensate as a highly saleable product. Its lightest components are separated and become LPG.

Tapti gas field

Offshore western India. Production was dropping, and information from logging at the several wells had indicated that the formation would over-respond to acid stimulation and become even more unconsolidated than it already was through the effects of drilling mud. Laboratory tests with core samples from the formation and various blends and compositions of acids were performed, and at one well successful stimulation was achieved with an acidising agent containing *inter alia* hydrofluoric acid.

TCP

Tube-conveyed perforating. (*See* **Perforating gun**.)

TDS

Total dissolved solids. (*See* **Raton Basin**.)

TEG

Triethylene glycol, a desiccating agent for natural gas. (*See also* **Bombay High North Field, glycol dehydration at**; **Glycol Dehydration**; **MEG**.)

Telemark Hub

In the Gulf of Mexico, comprising three oil fields: Mirage, Morgus and Telemark. The **pay** of Mirage and Morgus jointly is 81 m and the pay of Telemark is 43 m. Production and drilling are by means of the ATP[69] Titan, which is supported by three **spar platforms**. Notwithstanding its spar platfom support, ATP Titan does have pontoons for 'semi-submersion', so it transcends the definitions in this volume by being hybrid in its design. Drilling in the Telemark Hub has not been solely by the ATP Titan: **Ocean Victory** and West Sirius have also been in service there at the Mirage and Morgus fields. The initial activity of ATP Titan was production at Mirage and Morgus from wells already drilled. Drilling at the Telemark Field will be by ATP Titan.

Template

Device for securing a production facility, especially a **tension leg platform**, to the sea floor. As an example, at Shell's Auger tension leg platform in the Gulf of Mexico the templates comprise steel frames 18 m by 18 m by 9 m. There are four such templates each weighing 600 tonnes and they are secured by piles. The sea depth is 870 m. Production at the Auger platform is 25 000 barrels per day of oil with associated gas. A template can also be used instead of anchors in drilling operations, in which case it will be cemented into the sea-floor instead of being more 'irreversibly' attached by piles. (*See* **Shah Deniz**.)

Tengiz

Onshore oil field in Kazakhstan. Discovery was in 1979 and production began in 1993. The **pay** is 1.6 km. By 2004 production was 85 to 90 million barrels per year. At that time there were 53 wells, all of them 3500 m deep. In the early years of the 21st Century the field saw major development. This included sour gas injection (SGI): reinjection of associated gas high in sulphur in order to sustain reservoir pressure. This has raised the production to more than 150 million barrels per year. This is at a reservoir pressure of about 8000 p.s.i., and a *very rough estimate* of the **PI** can be obtained by using this instead of the drawdown pressure in the calculation. This follows.

$$\text{PI} = [150 \times 10^6 \text{ bbl year}^{-1}/(365 \text{ day year}^{-1})]/8000 \text{ p.s.i.} = 50 \text{ bbl day}^{-1} \text{ p.s.i.}^{-1}$$

which is within the range of those reported for very much more modestly producing fields in the entry for PI in this volume. A legitimate conclusion from a very rough calculation is that an exceptionally productive field need not have an unusually high PI. In addition to SGI, SGP – second generation production – has formed an expansion project. This is entirely downstream, being concerned with such matters as gas compression, refining and storage. (*See also* **Caspian pipeline consortium**; **Central Asia-China gas pipeline, Kashagan Field, Tolkyn Field**, **Zhanazhol Field**.)

Tension leg platform

Such an offshore oil production facility is held in place by tendons from the seabed to the platform providing stability and resistance to movement. The buoyancy forces on the hull of platform provide the tension in the tendons, which are held in place under the sea either by anchors or by piles driven into the seabed. A tension leg platform can be built onshore and towed on its hull to where it is required in the sea. It has a major advantage over the **fixed platform** in that it is suitable for sea depths of up to 2000 m. One of the world's largest production facilities to have a tension leg platform design is that at the Marco Polo Field in the Gulf of Mexico, which has been producing since 2004. The water depth is 1310 m. The hull of the platform was made by Samsung in Korea and weighs 5750 t and has a height of 60 m. It was towed to its destination in the Gulf of Mexico. There are eight tendons of 28 inch diameter. The piles attaching the platform to the seabed are 120 m long and of 2 m diameter. The Marco Polo Field produces 50 000 barrels of oil per day with large amounts of associated gas. Similarly, the Mars platform in the Gulf of Mexico (in the Mississippi Canyon, the scene of the Macondo Field where there was leakage over a period of several weeks in 2010) is a tension leg platform operated by Shell. It is in 900 m of water and produces of the order of 20 000 barrels of oil per day plus associated gas. (*See also* **Droshky Field**; **Kvitebjørn**; **Snorre Field**.)

Tern Field

Oil field in the North Sea, north-east of Shetland. The sandstone reservoir is of **porosity** >20% and permeability 350 md. The water depth is about 165 m. **Faults** in the structure are such that there is division of the field into two parts, with a mesurable difference in the depth of the oil-water interface: 2518 m TVSS (true vertical depth subsea) in the 'Main Area' and 2458 m TVSS in the separated area to the north. The platform at the field (Tern Alpha) has drilling as well as production facilities.

Terra Nova field

Offshore Newfoundland, scene of operation of a **disconnectable turret**.

The water is shallow, a mere 95 m. The disconnectable type was chosen because of hazards from drifting icebergs. Oil from the well having been received at the disconnectable turret (the hull of which is 'ice strengthened') is taken ashore by **shuttle tanker**.

Texas Eastern Transmission

One of the largest natural gas pipelines in the US, taking gas from Texas (where it is produced) to NYC. The gas therefore travels in a direction from south-west to north-east. In 2006 (the most recent year for which the author has been able to obtain the information) the throughput was 91 384 million cubic feet per day and there were seven **compressor stations** with a combined horse power of 1 517 465. Re-expressing in the units used in the entry for **Werne, Germany** this becomes 1.1 GW for the compression and:

$$(91384 \times 10^6 \times 0.028)\ \text{m}^3 \times 37 \times 10^6\ \text{J m}^{-3}/(24 \times 3600)\ \text{s} = 1.1\ \text{TW}$$

for the combustion, so the power required for gas compressor operation is a thousandth the hypothetical heat release rate for burning the gas at the compression rate explained in the entry for Werne. Of course, the device (turbine or engine) which is used at the compressor station will convert heat to work at an efficiency of only something like 35%, so the factor of one thousandth becomes:

$$(10^{-3}/0.35) \times 100\% = 0.3\%$$

which is still fairly insignificant. Losses through leakage even in the best maintained pipeline systems will enormously exceed this. The pipeline, because it is branched, has a significantly greater length than the shortest distance between its origin and its terminus, and the length of pipe is in fact 9200 miles. With 75 compressor stations this represents one every 122 miles, although simple division is probably too imprecise to give more than a very rough indication of the intervals between the compressor units when there is branching. (*See also* **GTN**, **South-East to North-East pipeline**, **Rio de Janeiro-Belo Horizonte Gas Pipeline II**, **Tuscarora pipeline**.)

TGP

Technip-Geoproduction. (*See also* **Harding Field**; **Shah Deniz**.)

Thamama

Group of oil fields in Abu Dhabi. One of them was the scene of successful use of **STEELSEAL®** to overcome **lost circulation**. It was used at 10 **ppb** level. A cellulosic bridging agent from the same manufacturer as STEEL-

SEAL® was also used, as was calcium carbonate. Clearly the **lost circulation** material was customised in this application. (*See* **SAFE-CARB**.)

Thebaud platform

Offshore Nova Scotia, and a major part of what is called the Sable[70] Gas Project. The Thebaud platform stands in 29 metres of water and is a **fixed platform**, prior drilling of the production wells having been by a **jack-up** rig. It has been producing natural gas since 1999. The operator is ExxonMobil. Other gas producing installations using a fixed platform in the Sable Project include Alma (water depth 67 m) and South venture (water depth 23 m). (*See* **Deep Panuke Field**.)

THERMO TONE™ 300

Chemical agent for **filtration** control manufactured by Halliburton. It is a polymeric substance, used in weights not exceeding 1% that of the base oil. Notwithstanding its polymeric nature it has no effect on mud viscosity if used in its intended temperature range, which is up to 150 °C. By contrast the sister product ADAPTA®, also a polymeric filtration control additive, is a 'secondary viscosity agent'. The origin of the difference might well be the degree of polymer cross linkage in ADAPTA®.

Throw

In drilling, the displacement due to a **fault**. (*See* **Meji Field**.)

Tiber Field

Those in the industry were excited to hear of the discovery of oil in the Tiber Field in the Gulf of Mexico in 2010. It was in very deep water, and it is believed that the field contains 3 billion barrels of oil. The discovery was dependent upon improved drilling and an account of the discovery on BP's web pages emphasises this and makes the following points in particular. R&D in drill bit manufacture is focused on use in 'harder-to-reach places'. Diamond drill bits have been used in the oil industry since 1863 and by now **PDC** is by far the most widely used diamond product in drilling for oil and gas. A good **ROP** is needed but (a point not touched on in the entry for ROP in this dictionary) so is a favourable cost per unit depth drilled. This requires drill bits with 'a longer lifespan'. Screw threads for drill bit attachment are becoming standardised so that a bit from one manufacturer can be replaced, in the long term or just for one phase of the drilling of a particular well, by one made by another. Drill bits can be rented on the basis of a price per unit length drilled and this is an increasingly attractive option for companies involved in exploration.

TIG welding

Tungsten inert gas welding, the name given to a process whereby a tungsten welding element is deployed in an inert gas such as argon ('shielding gas'). (*See also* **OTTO System**; **Underwater welding**.)

Tight gas, South American sources of

Information for the permeability of three tight gas fields in South American countries and the corresponding depths has been used by the author to calculate the conductivities from:

$$\text{conductivity (millidarcy-feet)} = \text{permeability (millidarcy)} \times \text{depth (feet)}$$

The calculated results are in tabular form below.

Field	Conductivity/md-feet
Caigu, Bolivia	53.5
Cerro Norte, Argentina	16.4
Llama, Venezuela	32.0

Previous discussions of conductivity in this volume have been for **proppants**, which are not part of the natural formation but introduced to enable hydrocarbon to pass through a scene of previous hydraulic fracture. The concept of conductivity obviously applies to a natural formation but comparisons of the conductivity values above with those given in another entry for proppants should be only on the broadest basis. (*See* **Southern North Sea, tight gas in**.)

Tog Mor

Pipe laying vessel in the AllSeas fleet. Lacking **dynamic positioning** it uses anchors for mooring, necessitating involvement of an anchor handling vessel in repositioning. It can lay pipes up to 60 inches in diameter. Like many other such vessels, Tog Mor[71] is also involved in laying risers. Clearly with a riser the most suitable procedure is **J-Lay**. (See **Solitaire**.)

Tolkyn Field

Gas field in western Kazakhstan also productive of condensate. Gas having been produced from wells at the field is taken to the Borankol oil field where **condensate stripping** takes place. The condensate is exported to Iran via a terminal at Opornaya, where there is a tank for condensate storage. Condensate production at Tolkyn is of the order of 6000 barrels per day.

Tombua-Landana

Development offshore Angola having recently commenced production. The water depth is 366 m, production is by means of a compliant tower the height of which is 474 m. Current oil production at Tombua-Landana is 100 000 barrels per day and there is a 'zero flaring' policy in relation to the associated gas. Similarly all 'produced water' is reinjected.

Toot Field

Oil field in Pakistan close to Islamabad, discovered in the early 1960s. Commencement of production was in 1967 and by 1986 there were seventeen production wells yielding a total of 2400 barrels per day of oil. The field is accessible by road and there is a pipeline taking crude oil from the field to a refinery. The **pay** is at depths of about 4500 m. Development work is now under way by the Canadian Sovereign Energy Corporation. Development will include the identification of natural fractures in the formation as sites for possible hydraulic fracture. (*See* **Salah Gas**.)

Top tension(ed) riser

Top tension(ed) risers are simply vertical, having no bends or other deviations in the geometry. They are used on the **tension leg platform** and on the **spar platform** and form a simple connection between the well and the offshore facility. Use is in deep water and the device will comprise multiple sections of pipe joined by threads or by welding. The top tension(ed) riser differs from the steel catenary riser (SCR), in which the pipe orientation is not vertical and attachment to the production facility is by a flexible joint. Such is in service at **Bonga**. The Auger platform referred to under **Template** uses the SCR. The SCR is very much a feature of the Gulf of Mexico and its first application elsewhere was in Brazilian waters at the Marlim Field.

Torm Kristina

A member of the same fleet as **Torm Ugland**, built in 1999. Torm Kristina has a deadweight of 99 999 tonnes and, like most of the Torm fleet (See **Torm Ugland**) carries refined products. In early 2011 Torm Kristina was the victim of piracy whilst off the coast of Oman. It was carrying gasoline. There was damage to the ship and injury to one of its occupants, but the pirates did not manage to board the ship which continued on its way to Mumbai. The sister vessel Torm Clara (deadweight 44 946 tonnes, year of build 2000) had been prey to pirates a week earlier whilst anchored close to the Malaysian coast. There was no harm to vessel or occupants. Charterers of Torm vessels include Shell, Statoil, Chevron and Total (in descending order of number of charters for 2009).

Torm Ugland

Vessels of the Torm fleet (HQ in Denmark) are used for refined products and condensate rather than for crude oil. Tom Ugland is, at the time of preparing this entry, in port at New York with a cargo of condensate. The deadweight of Torm Ugland is 73 400 tonnes, so an estimate, probably on the high side, of its payload of condensate is:

$$[(73400000 \text{ kg}/700 \text{ kg m}^{-3})]/0.159 \text{ m}^3 \text{ bbl}^{-1} = 0.66 \text{ million barrels}$$

This would of course have been **stabilised condensate**. In export figures condensate is always lumped with crude oil and to obtain a figure for the annual condensate import of the US has not proved possible. The only figure with which the figure above can be compared, though the basis for comparing them is tenuous, is with the annual *lease condensate* production in the US. 'Lease condensate' means condensate treated so as to remove its low boilers and, given the variety of practice which there is in the processing of such materials, can reasonably be taken to mean stabilised condensate. The lease condensate production in the US in 2009 was 178 million barrels. (*See* **Torm Kristina**.)

Toroa well

Offshore the Falkland Islands, an exploration well drilled from the **semi-submersible rig Ocean Guardian**. **Spudding in** took place in mid-2010. BHP Billiton and Falkland Oil and Gas Limited are the operators the former having 51%. The water depth is 600 m. Well drilling, to a depth of almost 2500 metres, was not rewarded by discovery of oil, so the well is now in the dry hole category.

Tortuosity, in well drilling

The reader is referred to the figure accompanying the entry for **AHD** and to the working in the box below.

The hypotenuse of the triangle in the figure is the measured depth (MD) and clearly:

$$\text{MD} = \text{TVD}/\cos\Theta$$

Continuing with numerical values for the BD-04A well offshore Qatar:

$$\text{MD} = \text{TVD}/\cos\Theta = (1.1/\cos 85°) \text{ km} = 12.6 \text{ km (41 338 feet)}$$

Now the 'drilling difficulty index' (**DDI**) is given by:

$$DDI = \log[(MD \times AHD/TVD)\tau]$$

with the lengths in feet, τ being the tortuosity. It is the number of degrees by which, along the drilling path, the drill bit will deviate from the idealised trajectory signified by the hypotenuse of the triangle in the figure. Now at the well under discussion DDI was 8.279, believed to be a record. Putting in the lengths gives:

$$\tau = 436 \text{ degrees}$$

This means that over the drilling distance the drill bit deviates from the idealised trajectory in total by 436 degrees in a directionally random way so that all such deviations are averaged out and the idealised trajectory represents the axis of the well when it is finally drilled. The calculation can be taken further in the following way. The curvature averaged over the trajectory is:

(436/126) degrees per hundred metres = 3.5 degrees per hundred metres or (more conventionally) 0.35 degrees per ten metres

and this value of the curvature is fairly typical. The very high value of the DDI at the field under discussion is due to the length.

The above definition of tortuosity is due to Schlumberger. Note that there is no reason whatsoever why the tortuosity should be limited to values up to 360° which indeed is exceeded in the above example. The matter of tortuosity and of DDI will be examined further using data for three wells offshore California. These are given in the table below, where the final column contains the calculated value for the angle Θ. It is possible to calculate the DDI for each of these if a generic value of 1 degree per ten metres is used in each case for the curvature.

Well name	Product	TVD/m	Measured depth (MD)/m	$\cos^{-1}$(TVD/MD)
State 3403 1	'Dry hole' well*	5079	5236	14 degrees
State C-5	Oil	4036	4339	22 degrees
SSMS 29208	Gas	3719	3837	14 degrees

*Not sufficiently productive for further development. (*See also* **Toroa well, Laos, oil exploration in**.)

The **pay** of State C-5 is 419 m, and that of SSMS 2920 8 is 70 m. The calculations are set out in the table below. The 'Schumberger equation' for DDI given above has been used to calculate the numerical values in the final column.

Well	AHD = TVD × tanΘ	τ = MD/10	DDI
State 3403 1	1266	524	5.83
State C-5	1753	434	5.91
SSMS 2920 8	927	384	5.56

The values of the DDI for the three wells are close to each other. If instead of 1 degree per ten metres for the curvature the value of 0.35 in the same units, calculated for the BD-04A well from knowledge of the DDI, is used, this changes the DDI by a factor of log(0.35) = −0.46 putting the values in the range to 5.10 to 5.45. That the value in excess of 8 for the BD-04A well *is* a record is not difficult to conceive. (*See also* **Boulton Field**; **Piceance Basin**; **TTRD**; **Woodford OK**.)

Tower mooring

A means of permanent attachment of an **FSO** or an **FPSO**. The tower has bearings at its base, enabling 'weather vaning' of the structure it is attached to. One example amongst very many of an FPSO so moored is the Bohai Shi Ji offshore China. In service since 2001, it can receive 80 000 barrels per day and can store up to one million barrels. Connection of the FPSO to the tower is by a wishbone (a.k.a. a 'soft yoke') whereby the apex of the triangular 'wishbone' is attached to the tower and the side of the 'wishbone' opposite the apex is attached the FPSO. Any movement of the combined structure originates at the bearings at the base of the tower. (*See also* **Kapap Natuna**; **Haiyangshiyou 102**.)

Trans-Alaska pipeline

Of length 800 miles, this pipeline conveys oil from Prudhoe Bay to Valdez. It was completed in 1977. Over about three quarters of its length it passes through permafrost terrain, which makes for difficulties with buried pipelines because of movement of the ground during freezing, which will impose stresses on any buried pipeline, resulting in its movement and jeopardy of its connection to other sections of pipeline. Similarly, thawing can cause such instability. Accordingly, parts of the Trans-Alaska pipeline are above ground and held in position by saddle supports, themselves secured by piles driven into the ground. These, being solid (unlike a pipe which is by definition hollow) can accommodate without movement the effects of freezing and thawing. In fact nearly half of the pipeline is so supported.

There are eleven **pumping stations** along the pipeline although they will not all necessarily be in use at any one time. With all pumping stations working and injection of a **drag reducing agent**, a pipeline throughput of the order of 2 million barrels per day is achievable.

Transocean Arctic

Semi-submersible rig, built by Mitsubishi and entering service in 1986. It can operate in up to 500 m of water and can drill to well depths of up to 7620 m. It has catenary moorings with eight lines each connected to an anchor weighing 15 tonnes. It received an SPS Overlay repair in October 2009. (*See* **Luno Field**, **Sedneth 701**.)

Transocean John Shaw

Semi-submersible rig having had long periods of service in the North Sea west of Shetland (although it carries the Panama flag). It can work in up to 500 m of water and can drill to 7620 m. Its **derrick** is rated at 635 tonnes. It is moored by eight lines each going to a separate anchor on the sea-floor.

Trent Field

Gas field in the UK sector of the North Sea, water depth 48 m. It was developed jointly with the Tyne Field and 'Trent and Tyne' is the usual name for the development, there being a platform at each. Gas from Tyne is sent to Trent for **condensate stripping** and gas from each field is sent by pipeline to Bacton off the Norfolk coast. At this stage operation of Trent and Tyne beyond 2012 is not assured. (*See also* **Pickerill Field**.)

Triethanol amine

$N(C_2H_4OH)_3$, one of a number of compounds which can be used as corrosion inhibitors to protect steel when a well is receiving acid stimulation. Dow Chemicals manufacture large amounts of triethanol amine for this application. Another is butyne diol – $OHCH_2C:C\text{-}CH_2OH$ – and another propargyl alcohol, HC:C-CH2OH. Korantin® PP (molecular formula $C_6H_{10}O_2$) and Lugalvan® P are chemical agents for the same purpose currently available from BASF. In oil field terminology the expression 'inhibited acid' often means hydrochloric acid containing such a reagent[72].

Tristan North West

North Sea field having undergone decommissioning in which the well intervention vessel Seawell was used as, at a later stage, was a **jack-up** rig from the Ensco fleet. A production well was plugged as was the 'discovery well' drilled in 1987. Each is now in 'plugged and abandoned' status. (*See also* **Ekofisk Field**; **Shelley Field**.)

Triton

Name given to a group of four oil fields in the North Sea: Bittern Field, Guillemot West Field and Guillemot North West Field. There is also associated gas, which receives **glycol dehydration** treatment. An **FPSO**, also called Triton, stands in Bittern Field at a water depth of 91 m. Wells from Guillemot West are tied back to it, and those from Guillemot North West are tied back to infrastructure at Guillemot West. Total production stands at 92 000 barrels per day of oil and 3.6 million cubic metres per day of associated gas. Oil is taken from the FPSO by **shuttle tanker** and gas by pipeline to an onshore terminal.

Triton™

Roller cone drill bit from Hughes Christensen having three rotating cones. The cutting surface is steel having previously been stress relieved by heat treatment.

Troll Field

An oil field with associated gas is by no means rare. One seldom encounters the term in reverse – 'gas field with associated oil' – but it would not be incorrect thus to describe the Troll Field, which is in the Norwegian Sector of the North Sea. The 2008 production figures are 138 000 barrels per day of oil and 120 million cubic metres of gas. A related calculation follows.

heat releasable from one day's gas/heat releasable from one day's oil

$$= 120 \times 10^6\ \text{m}^3 \times 37 \times 10^6\ \text{J m}^{-3}/(138\ 000\ \text{bbl} \times 0.159\ \text{m}^3\ \text{bbl}^{-1} \times 900\ \text{kg m}^{-3} \times 44 \times 10^6\ \text{J kg}^{-1})$$

= 5 to the nearest whole number.

When *pricing* is factored in, the half-an-order-of-magnitude higher figure for oil production will be eroded in that gas and oil do not sell equivalently on a heat basis, the latter being the more expensive sometimes by as much as about 40%. This makes the 'associated oil' able to raise more than a proportionate revenue.

A general point can be inferred from the calculation in the box above: where 'associated oil' occurs in small amounts there is stronger case for collecting it in preference to incinerating it than there is with associated gas. Water depths at Troll are about 330 m. There are three production platforms, Troll A, B and C. Troll A is a **GBS** and receives gas from 40 production wells. 'Troll B' is a semi-submersible and was a new build. It has been in service since 1995 and has catenary mooring. Troll C is also a semi-submersible – it has pontoons – but more commonly the acronym **FPU** is applied to it. The Troll Field accounts for over half of the gas from the Norwegian sector of the North Sea. There ought to be no difficulty in formal semantics or in operational practice distinguishing between associated oil and condensate: the former contains wax and the latter, of course, does not. Even so the term 'oil' is sometimes used synonymously with stripped condensate. (*See also* **Associated gas, value of**; **Oseberg Field**; **Kollsnes**; **Yttergryta Field**.)

Troy Williams

'**Slickline** lift boat' operated by Halliburton, actually a **jack-up** device capable of operating in water depths between 1.8 m and 21 m, i.e., in very shallow water. Its crane can lift up to 15 tonnes, which the reader might like to compare with weights of stuffing boxes across the range of sizes in which they are made. Troy Williams has two drums which can accommodate slickline in the diameter range 0.092 inches to 0.188 inches. Also operated by Halliburton is B.W. Hayley, a slickline boat having similar specifications to Troy Williams.

Truss spar

(*See* **Boomvang**.)

Trym Field

In the Norwegian sector of the North Sea close to the border with the Danish sector, productive of gas and condensate. The water depth is 65 m. It is tied back to the Harald platform 6 km away, which is actually in the Danish sector. The term 'cross-border tie-in' has been applied to it. Closure was on the agenda at Harald because of depletion of the reservoirs from which it received, so the 'cross-border tie-in', which is seen as a precedent for such arrangements in the future, has extended its life span and pushed back the expected date of decommissioning.

TT®

Smaller version of the **Hydra-Jar®**, from the same manufacturer. It is available in sizes up to 2⅞ inches o.d. This size has a detent force of up to 9 kN and a weight of 622 N. The ratio of the forces is therefore:

$$(9000/622) = 15$$

making the detent force an order of magnitude higher than the weight of the device. This is consistent with the inequalities deduced when the Hydra-Jar® was analysed in a previous entry.

TTRD

Through-tubing rotary drilling, a procedure under development by Statoil amongst other companies whereby a hole is drilled in the wall of an existing well and a branch well drilled from there. It obviously involves **directional drilling**. Everything at the parent well remains in situ, and this is the major advantage of TTRD. Its 'proving ground' has been the Åsgard field in the Norwegian sector of the North Sea. Well PH-4 at Åsgard had been developed by directional drilling and had a production life of only four years because of there being too much associated gas, which reduced the value of its yield. It shortly afterwards became the scene of TTRD trials. A **whipstock** was installed at a length of 3900 m along the well; because there had been directional drilling as noted, this was not of course a vertical depth. Conventional drill pipe attached section by section to the whipstock enabled the 'sidetrack' so created to extend another 1800 m in the direction determined by the angle between the whipstock and the original drill tube. There has also been successful application of TTRD at the Claymore Field in the UK sector of the North Sea and on the North Slope of Alaska. (*See also* **Åsgard pipeline**; **Kårstø processing plant**; **Njord Field**.)

Tuapse Trough

In the Black Sea, the scene of drilling for oil at water depths up to 2000 m. Rosneft and ExxonMobil are jointly involved in exploration over an area of 11 200 square kilometres. Also in the Black Sea is the Val Shatskogo project, in which Chevron have an interest. In a different part of the Black Sea, under Turkish jurisdiction, the newly commissioned **Deepwater Champion** will enter service in 2011.

Tuffcutter™

Range of **roller cone drill bits** from National Oilwell Varco, available in the size range 7⅞ inches to 12¼ inches. Its uniqueness is in the co-existence of **tungsten carbide** and steel in the cutting surfaces. Forging temperatures conventionally used in an all-steel bit will be around 1650°C, which would negate the effects of any tungsten carbide hypothetically present, as it would dissolve in the steel at such temperatures. With Tuffcutter™ forging temperatures do not exceed 1150°C, well below the temperature at which tungsten carbide dissolves. Tungsten carbide introduced before forging therefore retains its particle structure in the ultimate manufactured bit to the improvement of its performance in **ROP** and bit life terms.

Tungsten carbide

Chemical formula WC. It has a hardness equivalent to 9 on Mohs scale (ie. same as corundum, the indicator mineral). Tungsten carbide is used in drill bits for oil wells. Tungsten carbide is brittle and for the entire drill bit structure including the shank to be composed of it would make for breakages as well as being expensive. This difficulty is overcome by using a renewable tungsten carbide insert supported by steel. Tungsten carbide forms the analogue of an alloy with cobalt and with nickel. Such 'alloys' find application in drill bits for oil exploration in which case the nickel or cobalt is referred to as a binder. Both diamond and tungsten carbide are contained in a **PDC** drill bit. Tool steels are up to 8 on Mohs scale, below values for tungsten carbide or diamond (which is 10) but not having the brittleness of the former. Drill bits for CBM wells often have a tungsten carbide insert, as with the XPlorer (*See also* **Borrox®**; **BT1,2,3,3H**; **Drag bit**; **Drill collar**; **Matrix body**; **MX**; **PDC**; **PDW 6**; **Roller reamer**; **Silver Bullet**; **Spudding in**; **TuffcutterTM**; **Xplorer**.)

Tupi Field

(*See* **Lula Field**.)

Turboflo™

Drag reducing agent from Flowchem in the US. It is recorded on the company's web site that use of Turboflo™ in a particular crude oil pipeline network had enabled half of the **pumping stations** to be taken out of service. (*See* **Aging oil pipelines, use of drag reducing agents at**.)

Turret mooring

This comprises a turret to which mooring lines are connected on an **FPSO**. By reason of an assembly of bearings at the base of the turret, the hull of the vessel can rotate and be reorientated. The term often used is that the FPSO vessel can 'weather vane': the term 'swivel' is also used to describe the movement of an FPSO with turret mooring. Conventional catenary mooring lines, which have featured in several other entries in the dictionary, are used between the turret and the sea-floor. One advantage of the capability of the FPSO to 'weather vane' is that it makes for flexibility in the positioning of risers to the vessel. Limited weather vaning is provided for by **tower mooring**. (*See also* **Cossack Pioneer**; **External turret**; **Glitne Field**; **Norne**; **Jotun**; **Jubilee Field**; **Northern Endeavour**; **Terra Nova Field**; **Uisge Gorm**.)

Tuscarora pipeline

Whereas the **Texas Eastern Transmission** is one of the largest natural gas pipelines in the US, Tuscarora is a very small one taking gas only across the Oregon-Nevada border. It has one **compressor station** rated at 8900 h.p. (6.5 MW) and its throughput is 534 million cubic feet per day or:

$$[(534 \times 10^6 \times 0.028)\ m^3 \times 37 \times 10^6\ J\ m^{-3}/(24 \times 3600)\ s]\ W = 6.4\ GW$$

The factor of 10^3 linking gas throughput in thermal terms and compressor energy has been confirmed for Texas Eastern Transmission which is 9200 miles long and for Tuscarora which is 229 miles long, there being therefore one and a half orders of magnitude difference in the pipe length. Between Texas Eastern Transmission and Tuscarora there is a factor of seven difference in the number of compressor stations. The same factor occurred in calculations for the compression at **Werne, Germany**. There is scope for *cautious* general acceptance of it[73]. (*See* **South-East to North-East pipeline**.)

TVD

'True vertical depth', a very simple concept relevant to **directional drilling**. Where a well deviates from being vertical the length of the casing is not the same as the vertical distance from the top of the well to the level within the formation at which the casing either terminates or becomes completely horizontal. The latter is the TVD, and differs from estimates of the depth obtained by summing the lengths of casing tube installed which is sometimes called the measured depth. A related calculation follows in the shaded area below.

When drilling at the **Chayvo Field** was proposed a TVD of 2.6 km was aimed for with a horizontal displacement of 6 km to 10 km. If the drill casing is simplified to a straight line constituting the hypotenuse of a right angle triangle, the angle between the casing and the vertical is:

$$\tan^{-1}(8/2.6) = 72\ \text{degrees}$$

where the mid-range value of the horizontal displacement (8 km) has been used.

The calculated angle signifies the major horizontal component of the drill casing and therefore of the well. The gist of the calculation is without doubt correct, but it is a rough one intended to provide the reader with an illustration of the meaning of TVD by linking the definition to an oil field. Not only must the approximations be noted but so must the fact that there are several wells at the Chayvo Field. The term TD – total depth – refers to the depth at which there is **completion**, so conventional casing is replaced by casing with perforations. In the drilling at Chayvo the **ExxonMobil Fast-Drill Process** was used. (*See also* **AHD**; **Barrhead, Alberta, CBM at**; **DDI**; **Tern Field**; **Tortuosity, in well drilling**; **Tyra Field**.)

Tyra Field

Condensate field in shallow water (about 40 m) in the Danish sector of the North Sea. It was discovered in 1968 and entered production in 1984. By 1988 there were 36 wells and in 1989 the first horizontal well at Tyra was drilled. The horizontal part is 910 m long at a measured depth (MD) of 3050m giving a **TVD** of:

$$(3050^2 - 910^2)^{0.5} = 2900 \text{ m}$$

The field occupies a considerable area and **pay** varies in depth from 27 m to 67 m. There is a crest in the reservoir, and here the **porosity** is 40% to 50% and the permeability in the range 1 millidarcy to 10 millidarcys. At such permeabilities the gas is a very long way from being 'tight'. Current production of gas from the field is 6.7 million cubic metres per day.

UBCTD

Underbalanced coil tubing drilling. (*See* **Sidetracking, by coiled tubing drilling**.)

Uisge Gorm

FPSO in service at the Fife Field in the North Sea. It was a conversion, not a 'new build'. It operates in 72 m of water. It can receive 27 000 barrels of oil per day. It has **turret mooring** with nine catenary lines attached by piles to the sea-floor. It has facilities for **water injection** and for gas reinjection. (*See also* **Flora Field**; **Glas Dowr**; **Munin**.)

Ukraine, unconventional gas in

Under the heading 'unconventional gas' we include **CBM** and **shale natural gas**, each of which is being developed in eastern Ukraine at the present time by Eurogas Ukraine. The most potentially productive of the five concessions awarded to the company is Marijewvskogo Poligon, which contains shale natural gas in a quantity of 15.1 billion cubic metres. Horizontal drilling has begun there, by a Canadian drilling company, and Halliburton have a presence. As at Forest City, CBM and shale natural gas are in co-existence. (*See* **USA, selected CBM reserves in**.)

Umm Shaif Field

Oil field offshore the UAE. There are three **fixed platforms** at the field and oil production is 0.3 million barrels per day. There is associated gas and also, typically of fields in the Khuff formation, a reservoir of non-associated gas. The gas from Umm Shaif which is sent to the **ADGAS Plant** is associated gas. (*See also* **Bahrain Field**; **Bukhoosh Field**; **Yibal Field**.)

Under reaming[74]

To 'under ream' a well is to increase its diameter by means of a suitable downhole tool. Across the gamut of well engineering the term without qualification can mean more than one thing. A common meaning in the absence of further information is enlargement at the scene of **completion** to accommodate a gravel pack. (*See* **Metal Muncher™**.)

Underwater welding

The attraction of a device such as the clamp currently available for immediate use at the **Everest Field** is that it eliminates the need for subsea welding. Nevertheless, underwater welding is an important operation in the installation, maintenance and repair of offshore oil and gas facilities. A number of examples in tabular form are provided below. Comments follow the table.

Location and date	Details
Maui Field, offshore New Zealand, 1995.	Connection at a sea depth of 110 m of a 20 inch pipe between two platforms 15 km apart by TIG welding using the OTTO system. A plastic tent filled with argon around the welding site at each platform. This is said to be a 'first'.
Abu Dhabi, 2009.	Orbital welding (see below) tested for subsea conditions with positive results.
Offshore Brazil, present time.	TIG welding used in the installation of risers to an FPSO.
Norway, present time.	In situ repairs to immersed subsea power cables.

The reason for the 'plastic tent' at the Maui operation (row 1) was protection of pipeline's stainless steel lining, the purpose of which was to make the pipeline resistant to the sour gas which it had to convey. This was a thin lining in a pipe otherwise composed of carbon steel. Orbital welding (row 2) means welding around the entire circumference (that is, through an angle of 360°) of a static pipeline. In the oil field development which is the subject of the third row the **pipe laying vessels** are the 'sisters' Seven Oceans and Seven Seas, built in the Netherlands. The examples of subsea welding given above could be multiplied enormously. The examples in the table begin with an innovative one which those in the industry see as being state-of-the-art. It concludes with a much more routine one, the repair of 'umbilicals'. The discussion could have become more prosaic still by including repairs to **BOP**s. A reader will appreciate that subsea welding is both routinely conducted and the subject of new technologies. But that is true of anything, which is why we hear of new 'generations' of particular facilities and methodologies.

Urengoy Field

In western Siberia, productive of gas and condensate since 1977 and now operated by a subsidiary of Gazprom. Annual production is 260 million cubic metres of gas and 830 000 tonnes of natural gas liquids. The formation at the field is sandstone, and well depths are around 1200 m.

USA, selected CBM reserves in

The information in the table below is additional to that in the entries for **Raton Basin** and **San Juan Basin NM**.

Name and Location	Details
Powder River WY/MT	Producing **CBM** since 1989. > 30000 wells. Yield of gas reached one billion cubic feet per day in 2005. Up to 68 million barrels of water per month pumped out.
Uinta Basin, UT/CO	About 150 CBM wells. Methane **pay** at depths in the range 140 to 1400 metres.
Piceance Basin, CO	Formation at Piceance coal, sandstone and fine grained sedimentary rock. Coal seams within the structure targeted for CBM.
Black Warrior Basin, AL/MS	Producing since 1994. Now over 1000 wells.
Richmond Basin, VA	Large amounts of CBM in place in coal mines having ceased production in the 1920s.

Powder River is rated second only to the San Juan Basin NM as a CBM producer. At Powder River as at any scene of CBM occurrence produced water management is an important issue in expansion. The water-to-gas ratio for Uinta is higher than at San Juan Basin NM or at Raton Basin. The entry for Piceance Basin describes gas from parts of the field other than its coal seams, to which the CBM adds. In fact conventional gas (from the sandstone), **shale natural gas** (from the sedimentary rock) and CBM are all obtainable from Piceance Basin. Shale natural gas and CBM are both to be found at other basins including Forest City in KS. The latest USGS figures put the cumulative CBM production from Black Warrior at just under sixty billion cubic metres, indicating a time-averaged average daily production of:

$$6 \times 10^{10}/(17 \times 365) \text{ cubic metres per day} = 10 \text{ million cubic metres per day}$$

There has as yet been no commercial production of CBM at the Richmond Basin. There are many other CBM sources in the US which, like Richmond Basin, have been evaluated but have not entered production. It will suffice to mention one more: the Forest City Basin in eastern Kansas. This is seen as having promise, and the situation in 2011 is that existing wells there, having displayed a good 'show' of gas, that is, evidence of gas in the **drilling fluid** with the cuttings, are to be developed when pipelines for any gas produced have been installed. (*See also* **Barrhead, Alberta, CBM at**; **Fort Yukon AK, CBM at**; **Sekayu**; **Slimhole drilling**; **Ukraine, unconventional gas in**; **Vernon Field**; **Xplorer**.)

Utica natural gas shale field

In Quebec, the site of explorations and trials and now poised to enter production on a large scale. Companies involved include Canada's own Talisman whose horizontal well (additional to several vertical wells also drilled by Talsiman) at Utica is capable of producing 0.3 million cubic metres of gas per day. Utica is only 400 miles from New York City and that some of the gas will be taken there for use is highly probable. (*See* **Horton Bluff shale natural gas**.)

Utility pig

This term applies *inter alia* to a device carried along a pipeline by the liquid inside it, and occupying the entire cross section, in order to clean by scraping the interior surface of the pipe. This is also known as a cleaning pig and the **Bi-Di** range from Weatherford is an example. Pigging also has a role in **SRB** control. Any pig enters a pipeline at a *launcher* and exits it at a *receiver*. Note that the presence of a **butterfly valve** in a pipeline precludes use of a pig of any sort. (*See also* **Machar Field**; **Shah Deniz**.)

UWILD

Underwater inspection in lieu of drydocking. (*See also* **Gus Androes**; **David Tinsley**.)

Valemon Field

Gas field in the Norwegian sector of the North Sea containing a large proportion of condensate. Development is underway and commencement of production is expected in 2014. The water depth is 135 m. The appraisal well was drilled directionally from the **Kvitebjørn** platform and has a measured depth (MD) of 7380 m and a **TVD** of 4370 m. This makes the **AHD**:

$$[7380^2 - 4370^2]^{0.5}\ \text{m} = 5947\ \text{m}$$

Production wells will be drilled from the **semi-submersible rig Borgland Dolphin**. Once production has begun there will be some sharing of infrastructure with **Kvitebjørn**. (*See* **Njord Field**.)

ValuBond™

Resin coated **proppant** for use under conditions of temperatures up to 150 °C and closure pressures up to 8000 p.s.i., manufactured by Hexion in Cleburne TX. (*See* **Prime Plus™**.)

Vector pipeline

Whereas the **Alliance pipeline** and **GTN** convey natural gas from Canada to the US, the Vector pipeline conveys it from the US to Canada. Natural gas is admitted to the pipeline in Joliet Illinois, which delivers it to Ontario, offloading some on the way for use in Indiana and in Michigan. It has two **compressor stations** and can carry gas at up to 56 million cubic metres per day.

Vernon Field

Tight gas field in Louisiana. The permeability is in the range 0.005 millidarcy to 0.05 millidarcy across most of the field. One well there was recently the scene of hydraulic fracture. At this well the permeability was anomalously high – about 0.25 millidarcy – and Halliburton developed a high conductivity ceramic **proppant** for it. This comprised the proppant at 35 pounds per 1000 gallons of 'frac fluid'. The productivity of the well was increased by about half an order of magnitude.

Veslefrikk Field

Oil and gas field in the Norwegian sector of the North Sea, water depth 185 m. The reservoir depths are about 3000 m. Production began in 1989 and Statoil have a significant holding. The set-up for production is as follows. Oil and gas from 14 production wells go to a wellhead platform: there are also several wells for **water injection** or gas reinjection. Oil and gas from the wellhead platform ('Veslefrikk A') go to a **semi-submersible rig** ('Veslefrikk B') which is a rebuild and can receive 30 000 barrels per day of oil. Processing including **condensate stripping** takes place at Veslefrikk B. Oil and gas after separation are taken by pipeline respectively to the Sture terminal near Bergen and to Emden in Germany. (*See* **Shearwater Field**.)

Vesterled Pipeline

This carries gas from a field in the Norwegian sector of the North Sea to the terminal at St. Fergus, on the Scottish coast, by a link with the Frigg pipeline which begins at Alwyn North and passes close to the **Frigg Field**. Commencement of operation of Vesterled in 2001 was seen by many as being portentous. Norway's gas field operators had previously been unable to sell gas in the UK and commencement of such sale was naturally seen as signifying depletion of UK reserves, in particular at the Frigg Field.

Victoria gas field

(*See* **Ensco 100**.)

Virginia, CBM in

Virginia was producing coal before Independence in 1776 and there can be no question that in the mining of it then and more recently lives were lost through methane leakage into mine enclosures. One naturally expects that in this early 21st Century, when methane from coal mines is *desired*, Virginia would be 'proactive' and this is indeed so: **CBM** accounts for 80% of the methane production of Virginia and the state ranks fourth in CBM production. The principal CBM fields in Virginia are Oakwood, Nora and Middle Ridge. The table below gives some details of each.

Oakwood	Over 2000 wells operated by Consol Energy, the first having been drilled in 1984, and more being drilled. Hydraulic fracture used in the drilling of the majority. Recompletion planned at 30 wells in decline.

Nora	Concurrent production of CBM and **shale natural gas**. The latter is deeper.
Middle Ridge	Also operated by Consol. Hydraulic fracture. Vertical wells.

The three fields in the table are said to be amongst the most methane-laden coal fields in the 'western hemisphere', containing up to 600 cubic feet of CBM per US ton[75] of coal, a weight fraction of:

$$600\ \text{ft}^3 \times 0.028\ \text{m}^3\text{ft}^{-3} \times 40\ \text{mol m}^{-3} \times 0.016\ \text{kg mol}^{-1}$$
$$/[(1/1.1023)\ \text{tonne t}^{-1} \times 1000\ \text{kg tonne}^{-1}]$$
$$= 1.2\%$$

and that the CBM reserves are 1% of the weight of the coal reserves in a major coal-producing region is an indication of the abundance of the gas. On a heat basis this ratio becomes roughly:

$$(1.2 \times 55/30) = 2.0\ \%\ \text{to the nearest whole number.}$$

The quotient in the brackets is of course of the calorific value of methane divided by that of a typical bituminous coal, units in each case MJ kg^{-1}. A passer-by at, say, Oakwood might mistakenly think he or she was observing an oil field, as **pump jacks** will be evident. These are necessitated by the large amounts of produced water. If a CBM reserve is dry there is no need to make provision for water removal. The most important dry field, a very productive field poised for expansion, is that at **Horseshoe Canyon** in Alberta. At such a field nitrogen will be used in fracturing so as not to negate the benefits of the dryness. (*See* **Wales, exploration for CBM in**.)

Visund Field

Norwegian sector of the North Sea, producing oil and gas since 1999 and undergoing expansion. The water depth is 335 m. Drilling is by means of a **semi-submersible rig** configured as a **PDQ** facility. Statoil has a majority holding. Some of the associated gas is reinjected and there is provision for **water injection**. It is intended that production at Visund will be put on hold whilst two wells newly drilled in mid-2009 and very productive are tied back to it[76]. (*See* **Fault**; **Gullfaks**; **Kollsnes**.)

Wales, exploration for CBM in

In the entry concerned with **CBM** in Virginia a production figure of 600 cubic feet CBM per US ton of coal was given. This converts to 15 cubic metres per tonne of coal. In exploratory drilling for CBM in South Wales, values from three wells ranged across an order of magnitude from about 1 cubic metre per tonne to about 11 cubic metres per tonne. These were at permeabilities in the range 18 millidarcy to 44 millidarcy and the conclusion was drawn that the permeability has a major effect. (*See* **Airth**.)

Water injection

This is widely used as a means of aiding oil production from a reservoir. It is done at onshore and at offshore fields and requires the drilling of a **water injection** well. Crude oil and water are both incompressible, therefore the water injection and the improved oil production in volume units (conventionally barrels per day) will be about equal. It has been asserted that proposals to inject water at a rate of 12 million barrels per day into selected oil fields in Iraq will increase oil production from 2.5 million barrels per day to 12 million, an increase of 9.5 million barrels per day. The balance of about 20% is clearly due to loss of some water through its entry into parts of the formation away from the oil. The one-to-one 'rule' is therefore approximately upheld; it should be noted that the differing densities of crude oil and water are irrelevant to it. What is both relevant and important is the immiscibility of the two. Exxon are leading the operation referred to in Iraq. Seawater can of course be injected into an onshore field. An example *par excellence* of water injection is the Ghawar Field in Saudi Arabia, the world's largest oil field and entirely onshore. It receives several million barrels per day of seawater which is conveyed by a pipeline several hundreds of kilometres in length. (*See also* **Capixaba**; **Salinity, of produced water**; **SRB**; **Visund Field**.)

Watered out well

Term applied to a well which is producing at too low an oil-to-water ratio because of produced water (the effects of which can be worsened by injected water). The cut-off proportion of oil at which a well is deemed 'watered out' is arbitrary and a decision for field management. Such a decision at an offshore well will depend upon whether or not disposal of the produced water in the sea after oil separation is or is not permitted. (*See also* **Cook Inlet**; **Skua Field**; **Yibal Field**.)

Weaver's Cove MA

Scene of construction of a grass roots LNG terminal, completion expected in 2012. Target output is 120 million cubic metres of regasified product per day. Supply will be to consumers in MA, RI and CT. A small proportion of the LNG received at Weaver's Cove might be transferred by land transport to gas supply centres for use in peak shaving.

Well cuttings, effect of drilling mud on

Well cuttings are carried away from an oil well bore by the drilling mud. When the mud is water based the cuttings separate fairly readily and have the advantage of being free of oil unless some oil has been taken up from the well itself. If at an offshore facility they are returned to the sea they will disperse and do not contaminate the seawater. When the mud is oil based the cuttings obviously do retain some oil. Disposal at sea is then a less obvious option fraught with difficulties, not least that the oil acts as a cohesive agent and after admittance to the sea cuttings so disposed of can form a consolidated mass in the sea underneath the platform. The matter of cuttings disposability can be a factor in the making a choice between water based and oil based muds at offshore production platforms. (*See also* **Slimhole drilling**; **USA, selected CBM reserves in**.)

Well Enhancer

Semi-submersible rig, an intervention vessel with the capability to lower **slickline** into a subsea well. The operator is Well Ops. The vessel has **dynamic positioning** and two **moonpools**. Its building and commissioning were in response to an increased demand for well intervention as the number of aging wells increases with the passage of time and it has a reasonable claim to being the most advanced intervention vessel in service. From the same fleet is the well intervention vessel Seawell. This is in the Light Well Intervention (LWI) category has been in service in the North Sea for over 20 years. Its moonpool is 7 m × 5 m and it has dynamic positioning. Its **POB** is 135. Yet another member of the Well Ops fleet is the semi-submersible rig **Q 4000**, a well intervention vessel constructed in 2002. It operates in up to 3048 m of water and its POB is 133. Its moonpool dimensions are 11.9 m × 6.4 m.

Well stimulation vessel

Vessel from which hydraulic fracture can be carried out as can **matrix acidising**, by which is meant treatment of the geological formation with acid. (*See also* **Blue Angel**; **Blue Dolphin**; **Ceiba Field**, **DeepSTIM III**; **Singleton Field**; **Stim Star**.)

Welton

Scene of an oil field in Lincolnshire, England operated by Star Energy, about twice as productive as that at Humbly Grove also operated by Star Energy. The field is seen as having a life expectancy up to about 2016. Meanwhile Star Energy continue to explore in the region. (*See* **Albury**.)

Werne, Germany

Scene of a natural gas **compressor station** which is driven by a gas turbine the basis of which is a Pratt and Whitney jet aircraft engine of power 25 MW. The 'throughput' is 2 million cubic metres of natural gas per hour. This can also be expressed in watts by calculating the heat release rate were the gas *hypothetically* to be burnt at the rate at which it is compressed, and compressor performance is sometimes reported in this way. For the unit at Werne this would be:

$$(2 \times 10^6\ \mathrm{m}^3 \times 37 \times 10^6\ \mathrm{J\ m}^{-3}/3600\ \mathrm{s})\ \mathrm{W} = 20\ \mathrm{GW}$$

(*See* **GTN**.)

West Epsilon

Jack-up rig having seen its entire service in the various sectors of the North Sea. It can drill in water depths up to 140 m and to well depths of 9000 m. Its **derrick** can take a load of 900 tonnes and its **drawworks** are rated at 3000 h.p. It has a 5 K **blowout preventer** and a 15 K one. (*See* **Offshore Freedom**.)

West Navigator

Drill ship owned and operated by Seadrill, having seen service at scenes including the **Ormen Lange Field**, its location at the time of going to press. It can drill at water depths up to 2000 m and to well depths in excess of 10 000 m. It was built in Korea and entered service in 2000. It has **dynamic positioning** and its **derrick** can take a load of 680 tonnes. It is a competitive rig.

West Sole

Gas field in the North Sea in 28 m of water, discovered in 1965 and the very first gas field in the North Sea to be discovered. Drilling was from a **jack-up** rig called Sea Gem[77]. The formation is sandstone with the low permeability value of 12%. Production began in 1968 with two platforms 'A' and 'B', and the gas was taken to an onshore terminal at that time by pipeline. It required drying because of water in the formation: nowadays **zonal isolation** would most likely be applied to restrict the water from well entry. Further platforms followed and by 1978 there were 18 production wells to five platforms, one of which was taken out of service – the first decommissioning of a North Sea installation – that year. By 1980 the reservoir pressure had dropped and production was down. The current situation is that three nearby ('satellite') fields produce through West Sole facilities: Newsham, Hyde and Hoton. At Newsham and at Hoton there are sidetracks, and at Hyde there are three horizontal wells.

Westbury Micro-LNG Plant

LNG is a good automotive fuel. As was the case with liquefied petroleum gas and with methanol a few decades ago, consumer confidence and co-operation of suppliers at forecourt level will be required for its proliferation. BOC have set up a plant at Westbury in northern Tasmania which will convert natural gas originating in the Bass Strait to LNG at a rate of 50 tonnes per day for road use. There are four other such facilities strategically placed on the island so that the range of a vehicle on one tank will not exceed the distance between any two. Obviously a fairly sparsely populated island community like Tasmania with only one trunk route – from Hobart to the northern coast – is most suited to LNG vehicular usage. In the US and the UK the distance which can be travelled on a full tank of LNG and the distribution of service stations supplying it have been factors delaying the adoption of LNG automotive fuel. In such limited usage as there has been in the US evaporative losses from the tank have been shown not to be a major difficulty. (*See* **Bridgeport CT**.)

WGA-1

Like one having an organic base, an aqueous **hydraulic fracturing fluid** requires a gelling agent and Weatherford are amongst suppliers of such substances. One of their gelling agents for aqueous 'frac operations' is WGA-1, which contains a chemically modified form of the naturally occurring substance guar, a.k.a. guar gum. An aqueous hydraulic fracturing fluid will contain simple inorganics known as activators. Such inorganics suitable for use with WGA-1 include certain borates. The 'recipe' is 20 lb to 50 lb of the gelling agent to 1000 gallons of water and, consistently with a good frac fluid, the solid residue is low, 1% to 3%.

Whipstock

A section of drill tube the axis of which is inclined. The most obvious use is in directional drilling, where a whipstock enables two pieces of conventional drill tube to be connected at an angle (other than 180 degrees) to each other. When in drilling a drill bit has become detached use of a whipstock to divert drilling mud around the consequent blockage might be preferable to attempted retrieval of the bit. Note that the drill tube to which a whipstock is attached need not itself be vertical. (*See* **StarBurst™**, **TTRD** *inter alia*.)

WilPhoenix

Semi-submersible rig, built in Sweden and in service since 1982. It is now used in operations on existing wells rather than in the drilling of new wells. It can operate at sea depths up to 1800 m.

Window

Hole made in a well casing as the first stage in creating a sidetrack, requiring a **casing exit system**. (*See* **Metal Muncher™**.)

WOB

'Weight on bit', a factor in the performance of a drill bit. Such weight is of course provided via the drill structure. There has been much R&D into drill bits to give a low WOB and a good **ROP** especially in **directional drilling**. Obviously in directional drilling the gravitational forces will be modified from those in vertical drilling. The circulation of drilling mud is a factor in the effective WOB. The WOB is transferred to the drill bit from the **drill collar**, which is a tubular piece of steel (or some other alloy) through which drilling mud can flow. The drill collar and drill bit are components of the bottom hole assembly (BHA). (See also **Bit Booster®**; **Diamond EdgeTM**; **Down-hole motor**; **Drill bit, life expectancy of ('bit life')**; **Drilling at zero WOB**; **Hydra-Jar®**; **PWD**; **Secondary cutters**; **Shock absorber**; **Sledgehammer™**; **Sup-R-Jar®**; **Zhana Makat Field**.)

Woodford OK

The scene of **CBM** production, mid-2010 rate 0.58 million cubic metres per day. The operator is Petroquest Energy (HQ in LA). **Porosity** and permeability at the field are both low. Baker Hughes have had some involvement with developing Woodford, and this has included **extended reach drilling** to obtain lateral reach in order to access further **pay**. In this operation achieved **ROP** values averaged over the drilling period were 46 feet per hour and the steerable drilling enabled tortuosity to be minimised. (*See* **USA, selected CBM reserves in**.)

World Cup 2022

It was announced in late 2010 that the 2022 World Cup would be held in Qatar. There are already plans to obtain gas from the Qatar North Field – total reserves 25 trillion cubic metres – for electricity generation to service the World Cup and building projects ahead of it. Prior to the announcement there was a moratorium on increasing the amount of gas obtained from Qatar North which at the time of going to press is still in place. Qatar Petroleum have said that the moratorium will remain in force at least until 2014.

Wye piece

Section of pipeline shaped like a 'Y' enabling branching to occur. Often a newly laid pipeline will have wye piece incorporated with one branch blanked off but openable for future tie-in. (*See* **Nam Con Son pipeline**.)

X80 steel

For manufacture of pipeline for natural gas, the first such application in the US being at the **Cheyenne Plains pipeline**. It has a **yield strength** of around 550 MPa. This is higher than that for carbon steels commonly used in pipeline fabrication, and enables thinner walls to be used and hence the tonnage of steel for a pipeline to be reduced. The down side to X80 is that it is more susceptible to **hydrogen cracking** during welding than other widely used materials for pipelines. Stick electrodes for welding come in various compositions and are available in low-hydrogen forms. This was the choice for the welding of the pipes for Cheyenne. The second west-east pipeline in China is approaching completion at the time of going to press. It has a total length of 8660 km, comprising a main pipeline with branches. The main pipeline is made of X80 steel and will be the longest X80 pipeline in the world. Given the length of the pipeline one expects that the lower tonnage (and hence cost) of X80 to have been a factor in the choice. (*See* **Roma to Brisbane pipeline**.)

X-Bow (a.k.a. Axe Bow)

Design of ship such that the leading edge of the hull has a larger area and internal volume than in traditional designs. One of its earliest applications has been well intervention with **Sarah** and its 'twin sister' Karianne. The developer was the Norwegian Ulstein Group, who were involved in the construction of Sarah and Karianne.

Xikomba Field

Oil field offshore Angola, water depth 1480 m. The operator is Exxon. Production is by means of an **FPSO** also called Xikomba which was a conversion, not a new build. Transfer from it to shore is by **shuttle tanker**. At Xikomba there are four production wells, plus one for gas reinjection and four for **water injection**. Production began in 2004 and is currently 100 000 barrels per day. The field is in the same block as **Kizomba**.

Xplorer

Drill bit from Schlumberger having a **tungsten carbide** insert. An Xplorer drill bit of 4¾ inch diameter was used to drill two **CBM** wells in Pennsylvania where a depth of 4624 m was reached. Note that use of a bit of that diameter along the entire drilled length is a clear example of **slimhole drilling**. (*See also* **CBM Drill-In™**; **USA, selected CBM reserves in**.)

XR™

Reaming tool from Halliburton. The term 'eccentricity' as it applies to the drill strings is explained in the entry for **drill collar**. The claim that XR™ is a 'new generation' tool is on the basis that it incorporates features which promote *con*centricity thereby eliminating vibration.

Yadana Field

Gas field offshore Myanmar, discovered in 1980 (when Myanmar was Burma) and commencing production in 1998. The water depth is 40 m and the operator is Total. There are seven production wells gas from which is sent along a 650 km pipeline to an onshore site close to the Thai border. More than half of the gas is then directed to electricity generating plants in Thailand, and the undertaking is seen as a collaborative activity between the two Asian countries.

Yemen, LNG production in

The gas liquefaction plant in Yemen now has two LNG trains having for a considerable period had only a single one. The operator is Total. It takes gas from the Marib Fields and uses **MCR**®.

Yenangyaung

In Myanmar (Burma), scene of an oil field productive since the 19th Century. There was an attempt by Japan to appropriate the field in 1942. Much more recently – in 2008 and 2009 – development wells have been drilled there in a part of the field already having 20 producing wells. The development wells were drilled by Interra Resources, whose HQ is in Singapore. Interra are also active at the Chauk Field in central Myanmar, having re-entered the well Chauk 950 which was drilled in 1956 and 'shut in' in 1982. Twenty-eight years later, in 2010, production at the well was resumed. Other 'shut in' wells at Chauk have been so revived. Interra Resources are active in other countries in the region including Indonesia, where they have drilled at the Tanjung Miring oil field in South Sumatra.

Yibal Field

The largest oil field in Oman, reservoir depth 1420 m. Within the Khuff formation, it is operated by Shell. It began production in 1969 and soon afterwards became the scene of **water injection**. The reservoir is carbonate.

Its commencement predates horizontal drilling anywhere, but in more recent years there have been horizontal wells created at Yibal by way of horizontal sidetracks from existing vertical wells which had become watered out. In all 500 horizontal wells were drilled. **Zonal isolation** has also been applied. In 1997 the field was producing of the order of a quarter of a million barrels of oil per day. This had declined to 88 000 barrels per day by the year 2000 with amounts of produced water significantly exceeding that. There is gas at Yibal, but it is not associated gas: it is in reservoirs separate from the oil as at the **Umm Shaif Field**. During development at Yibal the absence of a **gas cap** was noted. The 2005 oil production figure for Yibal remained 88 000 barrels per day. The non-associated gas is conveyed 200 miles along the Yibal to Ghubra pipeline where it is used to generate power at about 500 MW. There is also a seawater desalination plant at Ghubra.

Yield strength

Property of metallic substances, the stress at which deformation begins. It has units equivalent to those of pressure. Yield strength is an important quantity in the metallurgy of oil and gas production. The range of yield strength of carbon steels used in well casing encompasses 55 000 p.s.i. and 110 000 p.s.i., two widely used values. A value of 55 000p.s.i. converts to 380 MPa (or equivalently N mm^{-2}) and a value of 110 000 p.s.i. converts to 780 MPa. Examination of this quantity for the drilling jar follows.

In examining the **Hydra-Jar®** the 6½ inch o.d. size was used. Its i.d. is 2¾ inches, giving the pipe annulus an area of:

$$\pi(6.5^2 - 2.75^2)/4 \text{ square inches} = 27 \text{ square inches} = 0.0175 \text{ m}^2$$

The tensile yield strength of this size of Hydra-Jar® is given as 900 000 lb = 4 004 082 N

Dividing one quantity by the other gives:

$$(4\,004\,082/0.0175) \times 10^{-6} \text{ MPa} = 57 \text{ MPa}$$

This is about an order of magnitude lower than generic values given previously in this entry, but it appertains to use at 200 °C or higher whereas the previous values are for ordinary temperatures. This is probably the origin of the difference. The values for aluminium and for copper at ordinary temperatures are below 100 MPa.

(See **X80 steel.**)

Yme Field

Offshore Norway, water depth 93 m. Productive of oil from 1996 to 2001 with Statoil as operator, yielding 50 million barrels over this period. Production is set to resume with Talisman as operator and further wells are being drilled by means of a **jack-up** structure. There will be wells for **water injection** and for gas reinjection as well as production wells. (*See* **Oseberg Field**.)

Yoho Field

Offshore Nigeria, water depth 60 m to 90 m. The reservoir depths below the seabed are also in this range. Production is at an **FPSO** ('FPSO Falcon') which was a conversion from a tanker. It is moored by an **external turret** to a buoy and has **water injection** and gas reinjection capabilities. Current production is 90 000 barrels of crude oil per day.

Yolla-A platform

In the Bass Strait of southern Australia, a **GBS** built by Arup. The water depth is 80 m. It receives gas and condensate from the Yolla Field. (*See also* **Hanze F2A Field**; **Maari Wellhead Platform**.)

Yttergryta Field

Gas field in the Norwegian sector of the North Sea, water depth 300 m. It was discovered in 2007 and began producing in 2009. The reservoir depth is 2390 m. Gas from a single well at Yttergryta goes to an **FPU** which, like the one at **Troll Field**, has pontoons and receives from three other gas fields: Midgard, Smorbukk and Smorbukk Sor. At Yttergryta there was **J-Lay** by the then newly commissioned pipe laying vessel Seven Seas[78]. At the FPU 36 million cubic metres of gas and 94 000 barrels of condensate are produced daily. The weight ratio of condensate to gas is approximately:

$$94\,000 \text{ bbl} \times 0.159 \text{ m}^3 \text{ bbl}^{-1} \times 700 \text{ kg m}^{-3}/(36 \times 10^6 \text{ m}^3 \times 40 \text{ mol m}^{-3} \times 0.016 \text{ kg mol}^{-1}) = 0.45\ (45\%)$$

The same calculation using data given for the Dolphin Gas Project, Qatar gives a value of 31%. (*See also* **Angel Field**; **Condensate stripping**; **Markham Field**; **Platong II Project**.)

Yùum K'ak' Náab[79]

FPSO in service in the Gulf of Mexico, an **STP** facility. It is an example of the use of STP at a part of the Gulf – the Bay of Campeche – very susceptible to hurricanes. Made in Singapore by conversion of a tanker initially constructed by Mitsui, Yùum K'ak' Náab can hold 2.4 million barrels of oil and operates in 100 m of water.

Z

Z-16 ST

Well at Sakhalin, the scene of a successful sidetrack operation necessitated by a 'fish' (detached drill bit). The 'fish' was at a measured depth (MD) of 8197 m and the sidetrack was made at 7800 m. (See also **Borrox®**; **HCR 506ZX**.)

Zaafarana Field

In the Gulf of Suez, water depth 57 m. The field was discovered in 1990. There is an unmanned wellhead platform attached to the seabed by piles. Oil from this goes to an **FPSO** which is permanently moored. This recently received repairs by SPS Overlay, avoiding the need for it to go to a yard for repair with all that that would have involved in terms of disruption to production. (*See* **Conkouati**, **Sedneth 701**.)

Zeepipe I

Pipeline carrying gas produced at Sleipner, in the Norwegian sector of the North Sea, to a terminal at Zeebrugge in Belgium, operating since 1997. It was followed by Zeepipe II and Zeepipe IIA which receive gas from **Kollsnes**. Major pipelines for gas from Norwegian waters meriting a mention but not a separate entry are *Franpipe*, which takes gas from an offshore platform to the French coast and *Europipe*, which takes gas from the same platform to a terminal Germany. *Condensate* from Sleipner is taken to the **Kårstø processing plant** by the Sleipner Condensate Pipeline, which is of length 245 m and entered service in 1993. It is operated by Statoil. Fabricated of 20 inch diameter pipe it can convey 200 000 barrels of condensate per day. (*See* **Statpipe**.)

Zhana Makat Field

In Kazakhstan, undergoing exploration and development concurrently with initial production. The operator is Max Petroleum, who report on their web pages that over the period April to July 2010 average daily production was

2100 barrels. The formation is sandstone of **porosity** 18% to 26%, and oil **pay** is at depths between 1282 m and 1410 m. The drilling rig in use at Zhana Makat is owned by the Chinese company Sun Drilling, in which Sinopec has a stake. It is model ZJ30DBS in which the **WOB** adjusts with conditions and the remaining bit life can be judged. With 5 inch drill pipe the depth limit of the ZJ30DBS is 3000m. The **drawworks** are rated at 600 h.p. and the height of the **derrick** is 41 m.

Zhanazhol Field

Onshore oil field in Kazakhstan, discovered in 1960[80] and producing since 1987. Reservoir depths are of the order of 5 km and in places the permeability is as high as 2 darcys, **porosity** being 20% to 25%. Current production is about 0.1 million barrels of oil per day. (*See also* **Kashagan Field**; **Kazakhstan-China oil pipeline**; **Tengiz**; **Tolkyn Field**; **Zhana Makat Field**.)

Zonal isolation

(*See* **swellable elastomer**.)

Zuata Field

Oil field in Venezuela, within the Orinoco Belt, oil from which is of density 9 degrees API (density at 60 °F = 1000 kg m^{-3}). An **electric submersible pump** is in service there. Its production, like that of so much Venezuelan oil, requires use of a diluent. Recent production figures at Zuata are 2240 barrels per day with injection of 560 barrels per day of recoverable diluent. In general naphtha from refining is suitable for diluent use. The diluent is also used in piping oil a distance of 120 miles to an industrial site and after separation is returned, also by pipeline, to Zuata. The pipe conveying the diluted oil is of 36 inch diameter so has a capacity of about 20 million barrels (a simple calculation which the interested reader can easily repeat). Initial filling took about four weeks. The pipe conveying the separated diluent is of 20 inch diameter. At one of the other oil fields in the Orinoco Belt downhole heating is being used as an alternative to dilution. This is at a horizontal well of 1000 m length at a **TVD** of 1219 m. The result of application of heat at 65 kW was, once approximately steady thermal conditions had been reached, a 65% rise in the production at the well. (*See also* **Drill bit, life expectancy of ('bit life')**; **LL652 Field**; **Mesa 30**; **Rajasthan Block**.)

Postscript

I have often over the years wondered whether the words 'engineer' and 'ingenuity' are etymologically linked and have intuitively believed that they probably are. If I rewrite them as e***ngin***eer and i***ngen***uity it is clear that the consonant structures of the respective words are consistent with such a link. My web research has informed me that 'engineer' comes from the Latin *ingeniare* which means to 'devise'. It would be a waste of time for someone like myself lacking linguistic training to attempt to pursue the research any further, though I'm sure that amongst users of the book there will be those who could clarify this point for me. I did think of seeking guidance from linguists or classicists at the University of Aberdeen on these matters and have no doubt that it would have been willingly given, but decided instead not to digress further from the subject of the book.

I have begun the postscript on this note because time and again on preparing material for this book I have been deeply impressed by the *ingenuity* displayed in oil and gas production matters. I have tried to communicate these impressions in the entries by giving detailed accounts of operations and practices. Comprehensiveness has been aimed for in this volume and the arrangement is alphabetical but (a point which I touch on in the Preface) the content goes deeper than might be expected of a dictionary. Indeed, someone attempting to write a 'dictionary' of oil and gas production comprising mere definitions would soon encounter the difficulty of ambiguity and uncertainty in such 'definitions' and of regional variations in word usage. That being said, the 'glossaries' of terms in oil and gas production which there are on the web *are* useful and have indeed been of help to me. These however do not give bald 'definitions' but solid information with links to related entries accessible by a simple click.

These are exciting times for anyone interested in oil and gas production. When in my own thoughts I review the oil industry from its beginnings to recent times an expression from Holy Writ comes to mind: We have a 'goodly heritage'[81].

Endnotes

1 On 14 April 2011 Walter Breuning of Montana USA died at age 114. In the press accounts it is recorded that he recalled having been given first-hand accounts of the Civil War by his grandfather. The dates usually given for the American Civil War are 1861 to 1865. These dates lie between the drilling of Drake Well (1859, in PA as noted) and the foundation of the Standard Oil Company by J.D. Rockefeller (1870, in OH), so these events would have been during the adulthood of Bruening's grandfather. Rockefeller himself was in fact conscripted for the Civil War but did not fight in it as his offer to pay for a substitute (a practice which was also common in the Napoleonic wars) was accepted.

2 The 2011 price of natural gas was about $1 per thousand cubic feet.

3 Libya is in a state of emergency at the time of writing this entry (November 2011). (See also **Torm Ugland**.)

4 The term 'multiple reservoirs' is obviously correct here, and sometimes the term 'multizone reservoir' (note the singular) is used synonymously. The author's preference is for 'multiple reservoirs'. The term 'multizone stimulation', meaning measures taken to increase oil production from such a formation, is unexceptionable.

5 http://www.searchanddiscovery.net/abstracts/html/2008/intl_capetown/abstracts/471404.htm

6 This field is at an early stage of production: oil production is expected to have risen by an order of magnitude by about 2015. Whether gas production will rise proportionately will depend inter alia on the production practices followed, and is indeed a factor relevant to the 'value of associated gas'.

7 Alternatively, Moulavi Bazar.

8 Jakarta Globe, 18 April 2011.

9 Bi-Di means bi-directional: the pig can be sent in one direction or the other along a pipeline. Weatherford applies the term to its products but as far as the author has been able to ascertain it is not a trade mark.

10 In all such matters an accepted meaning has to be just that, accepted by a user without undue preoccupation with alternative meanings if taken at face value. Nobody would dispute the validity of the term 'well stimulation', which occurs many times in this volume, but is it not the reservoir which is really experiencing

'stimulation'? Or is it that at the site of **completion** 'well' and 'reservoir' are not distinguished? (See also **Acid fracturing pumper**.)

11 There is also a 'Bruce platform' at **Cook Inlet**.

12 Described in one coverage as a 'web of thin reservoirs'.

13 The text 'A Cubic Mile of Oil' by H. Crane, E. Kinderman and R. Malhotra was published by OUP in 2010.

14 In a clamp the tightening action is in a direction perpendicular to the axis.

15 API density is of course on an inverse scale.

16 'Sandwich Plate System'

17 A photograph of it is to be found at: http://www.alaska-in-pictures.com/monopod-oil-platform-3221-pictures.htm

18 bbl = barrel

19 http://www.cairnenergy.com/NewsDetail.aspx?id=1342

20 For a photograph of which the reader should go to: http://www.oilrig-photos.com/picture/number1802.asp

21 The acronym 'SBM' has another meaning in offshore oil production: 'single buoy mooring'. An **FPSO** or an **FSO** can have single-buoy mooring as can a tanker. The FPSO in use at **Yoho Field** has single-buoy mooring.

22 Even so, a **semi-submersible rig** installed for an indefinite period at a particular scene of production might be referred to as a 'platform', as at **Veslefrikk Field**.

23 Compliant = flexible in this sense.

24 International Maritime Organization.

25 This manufacturer supplies TNK-BP, which carried out the hydraulic fracture at **Severo-Varyoganskoe** amongst many other places.

26 In a direction orthogonal to the pay as measured.

27 **GBS** is sometimes intended to mean gravity based substructure, referring to parts of the platform excluding the topside and (if one is in service) the caisson. It is doubtful whether there is any difference between the two interpretations of the acronym.

28 'Red Dragon' in the local language.

29 The following is from an SPE publication from 2000: In 1995, the first horizontal sidetrack successfully drilled using coiled tubing (CT) was carried out at Shell's House Mountain oilfield. The objective was to drill underbalanced a 3⅞ inch diameter by 300-m-long horizontal hole section from an existing vertical, cased oil well used to directionally drill the horizontal section underbalanced with nitrified water.

30 See 'The Times' newspaper, 25 March 2011, p.18. There it is stated that Valiant employs only 20 persons but 'provides work for scores more' as contractors and indirectly in its co-operation with bigger companies and its infrastructure sharing.

31 It is interesting to note that this exceeds the nameplate depth limit of 100 m given in the entry for **Maersk Gallant**. It is a matter of judgement whether limits

specified by the designer and builder can in particular circumstances be adjusted and if so by how much. Another way of saying that is that there is some flexibility in the specification.

32 In judging and evaluating the magnitude of the detent force above, let it be noted that the thrust of an aircraft jet engine will be of the order of tens of kilonewtons.

33 Azipod® is a registered trademark of the ABB Group.

34 At the **Map Ta Phut LNG Regasification Terminal** the dredging work carried out by a **cutter section dredger** was in fact about five million cubic metres. It was reported on the Web at the time that this operation began that 'the [dredging] project started in January 2008 and is scheduled to be completed in April 2009'. Approximating then the time to 12 weeks, the rate of dredging was:

$$(5/12)\text{ million cubic metres per week} = 0.4\text{ million cubic metres per week}$$

Van Oord have twelve **cutter section dredgers** in their fleet, the most powerful of which in terms of cutting capability entered service in 2011 so could not have been used at Map Ta Phut. The author has been unable to ascertain which of the other eleven was used. Some of the smaller ones are not self-propelled.

35 One coverage has to say of this: *Shell's move is the latest sign of the major's attempt to replicate the success of the North American unconventional gas industry around the world.*

36 A misnomer: *see* **Sunkar floating production vessel**.

37 http://www.alaskajournal.com/stories/021011/loc_klpc.shtml

38 http://www.onepetro.org/mslib/servlet/onepetropreview?id=00026334&soc=SPE

39 Note the explanation in the entry for **Troll Field** why unit heat from oil products and natural gas are not equivalent when expressed in monetary terms.

40 Though a founding member of OPEC, Iraq has not of course been assigned an OPEC quota for about the last 10 years. That was in the author's view a factor in the admission of Angola and readmission of Ecuador to OPEC, both in 2007.

41 A trade mark of Air Products and Chemicals Inc. (APCI). Natural gas production plants utilising such refrigerants are often described as using 'APCI technology'. (*See also* **Damietta.**)

42 There is a delightful quip appertaining to the lack of oil in Israel which is often associated with former Israel PM Golda Meir, although it is doubtful whether she was the originator of it: 'Why did Moses lead us to the one place in the Middle East without oil?'

43 A brine in the sense of that term in chemical engineering is not always (or even most commonly) seawater. It might be an aqueous solution of a salt such as calcium chloride or, if pipe corrosion by an ionic solution is a difficulty, of ethylene glycol. At the refinery at Mizushima the 'brine' *is* seawater but the term 'brine' could be applied as accurately to one of the 'synthetic' cooling fluids described in the previous sentence.

44 This is of course a simple example of 'process integration'.

45 As is so often the case there is some inconsistency in terminology here. The term 'pipeline sphere pig' is an accepted one but such a device might simply be called a 'sphere' and distinguished from a 'pig' which, one has to assume, is on that definition any shape other than spherical! A sphere will have a soaking rather than a scraping action and this might in the minds of some in the industry be the basis for the distinction. A 'sphere' in this sense might be solid or a softer, sponge-like material swollen by uptake of water or a glycol preparation.

46 National Oilwell Varco also have a product called Near-Bit Reamer (note the hyphen).

47 It is perhaps worth a comment that Abu Madi produces about twice as much natural gas liquids (NGL) as Al-Qar'a but the yields in the LPG carbon number range are the same. Subtle points like this can affect the viability of NGL processing.

48 On 5th February 2011 there was an explosion at this pipeline in Sinai receiving gas from Port Said. There were no injuries or deaths. Supplies from Port Said to other countries including Israel have been affected.

49 The O-ring was invented by Niels Christensen and a patent was granted in 1939. It is a point of some irony that Edison's 1882 light bulb patent application shows a rubber seal where the metal and glass of the light bulb are in contact. This was arguably an unintentional anticipation of Christensen's invention.

50 http://www.spe.org/ejournals/jsp/journalapp.jsp?pageType=Preview&jid=EDC&mid=SPE-114008-PA

51 Crude oils denser than water are rare.

52 General definition of caisson, obtained from the web pages of a manufacturer: 'a watertight box for underwater construction work'. (*See also* **Nini and Cecilie Fields**.)

53 It is said that the rather whimsical term 'pig' has its origins in the fact that such a device on progressing along a pipe creates a squealing sound!

54 For a striking photograph of which the reader can go to: http://www.gazprom.com/production/projects/deposits/pnm/

55 The Maori word for 'old age'. However, RAROA is also a recognised acronym meaning 'Risk-Adjusted Return On Assets'. It is difficult to know which meaning, if either, was in the mind of whoever so named the facility.

56 Part of the Douala Basin including **Logbaba** is *on*shore.

57 This claim had, prior to its tragic demise, been made on behalf of **Rocknes**.

58 During a hydraulic fracturing operation at a well in the part of Marcellus in PA ('Atgas 2H well') by Chesapeake in mid-April 2011 there was loss of containment of fracture fluid which consequently entered the environment in quantities of thousands of gallons.

59 There is a view, not necessarily the ultimately authoritative one, that the fundamental distinction is in the permeability: up to 0.001 millidarcy for shale gas and up to 1 millidarcy for tight gas, irrespective of the nature of the formation. Not a few conventional gas fields qualify for 'tight' classification on this

criterion. The definition probably reflects the fact that permeabilities typical of many shale natural gas reserves would not be expected at conventional fields. Approximately synonymous use of the terms represents current usage in the industry. The permeability figure for tight gas in this footnote is lower than that given for Green River WY but not by an order of magnitude. (See also Argentina, unconventional gas plays in, Woodford OK.)

60 The same manufacturer offers a range of **downhole motors** of power up to about 10 h.p.

61 Each from from the Teekay web pages.

62 There is also a Skua Field offshore Australia.

63 An SPE conference paper on Brage describes the formation as being 'streaky'.

64 In the entry for Texas East Transmission, and those to which it is linked.

65 The following is taken from a discourse on natural gas on the web: *Typical conventional natural gas deposits boast a permeability level of .01 to .5 darcy, but the formations trapping tight gas reserves portray permeability levels of merely a fraction of that, measuring in the millidarcy or microdarcy range.*

The low end of the range given in the main entry converts to 10 microdarcys. (See also Vernon Field.)

66 For a photograph of one of these four spud cans go to: http://www.lamprell.com/lamprell/media/newsletter/spring_nl_2009/spring_nl_2009.pdf

67 This is an accepted acronym. Occasionally the same letters are used casually to denote semi-submersible drilling vessel.

68 'Falcon' in the local language.

69 Denoting ATP Oil and Gas Corporation, founded 1991 with its HQ in Houston.

70 There is also a Sable *Field* offshore South Africa. (See **Glas Dowr**.)

71 For a photograph of **Tog Mor** showing the stinger in the downward position go to: http://www.allseas.com/uk/59/equipment/tog-mor.html

72 The term 'inhibited acid' should not be confused with 'retarded acid', which means acid containing a thickener to control the rate of diffusion into the formation. Hydrogen fluoride, in view of its low molecular weight, often needs to be so 'retarded' when used in well stimulation. The term SRH-RHF – slow-release-hydrogen retarded-hydrofluoric-acid – has been coined by a major well services company.

73 A reader disposed to do so has enough information to repeat the calculation for other pipelines including Alliance and **GTN**.

74 UR™, denoting a tool for under reaming, is a trademark of Halliburton.

75 1 tonne = 1.1023 US ton

76 On Saturday 9 April 2011 there was a gas leak at the Visund field. Occupants of the platform were evacuated and production suspended until operator Statoil had had the opportunity to investigate.

77 This has a sad dimension. Having made the discovery of gas described Sea Gem was being dismantled in readiness for movement to another scene of activity

when it capsized with loss of 13 lives. The other 19 occupants were rescued.

78 Not to be confused with the drill ship 'Discoverer Seven Seas', which features in two entries of this dictionary.

79 'Lord of the Sea'

80 When Kazakhstan was Kazakh SSR.

81 Psalm 16 verse 6, King James Version.